Aquatopia

Aquatopia documents Harmattan Theater's ecological interventions and traces its engagements with water-bound landscapes, colonial histories, climate change, and public space across New York City, Venice, Amsterdam, Lisbon, and Cochin. The volume uses Harmattan's site-specific performances as a point of departure to consider climate change and rising sea levels as geographical, ecological, and urban phenomena. Instead of a collection of flat, static surfaces, the *Aquatopia* atlas is animated by a disorienting, anti-mapping strategy, producing a deterritorialized, nomadic, fluid atlas unfolding in real time as an archive of climate change in multidimensional, active space. The book is designed for pedagogical access, with interludes that consolidate the learning outcomes of the experimental theory animating each site-specific performance.

Accompanied by close descriptions of five performances and supplemented by digital documentation available online, this volume intervenes in discussions on climate change, urbanism, and postcolonization/decolonialization, and contributes to interdisciplinary studies of ecology and environmental politics, postcolonial/decolonial theories and practices, performance studies and aesthetics, in particular public art, and performance as research.

May Joseph is Founder of Harmattan Theater, Professor of Social Science at the Pratt Institute, and author of the *ghosts of lumumba*; *Sealog: Indian Ocean to New York*; *Fluid New York: Cosmopolitan Urbanism and the Green Imagination*; and *Nomadic Identities: The Performance of Citizenship*. Joseph is co-editor of *Terra Aqua: The Amphibious Lifeworlds of Coastal and Maritime South Asia* and of *Performing Hybridity*. She co-edits three book series from Routledge: *Critical Climate Studies*, *Ocean and Island Studies*, and *Kaleidoscope: Ethnography, Art, Architecture and Archaeology*. Joseph creates site-specific performances along Dutch and Portuguese maritime routes exploring climate issues. Visit www.mayjoseph.com.

Sofia Varino is a writer and public scholar whose work focuses on radical thought and practice, cutting across political ecology, history and philosophy of science, and transdisciplinary gender studies. They have published in journals like *Whatever, SHIMA, European Journal of Women's Studies*, and *Women's Studies Quarterly*, and co-edited a special issue of *Somatechnics* on data and gender in the life sciences. Varino is a postdoctoral researcher affiliated with the minor cosmopolitanisms research training group, a cooperation established among the University of Potsdam, Humboldt Universität zu Berlin, and Freie Universität Berlin in Germany. Visit sofiavarino.com.

Critical Climate Studies

The Critical Climate Studies book series is located in the transdisciplinary space that crosscuts the social sciences, humanities, creative writing, environmental studies, and climate science. Scholarship and activism are powerful but often invisible global forces, trapped in the interstices. We seek to draw attention and analysis to such domains. The series welcomes short books that experiment with holistic engagement, critique, and conversation about climate change, broadly conceived. In addition to nuanced academic prose from all disciplines, the series embraces multi-genre writing, experimental ethnographies, creative non-fiction, lyrical sociology, ficto-critical writing, as well as science-humanities collaborations. We encourage contributions that are investigative, immersive, and attentive to the understudied and obscured planetary transformations taking hold as climate change accelerates. Our interests lie in large debates as well as in the understudied regions and microhistories of the world, where the impact of the planet's climate convulsions generate altered experiences and analyses of ontologies, geographies, ecologies, and political economies.

Barbuda
Changing Times, Changing Tides
Edited by Sophia Perdikaris and Rebecca Boger

Aquatopia
Climate Interventions
May Joseph and Sofia Varino

For more information about this series, please visit: https://www.routledge.com/Critical-Climate-Studies/book-series/CCS

Aquatopia

Climate Interventions

May Joseph and Sofia Varino

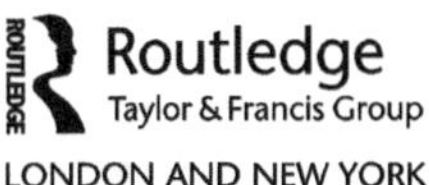

LONDON AND NEW YORK

First published 2023
by Routledge
4 Park Square, Milton Park, Abingdon, Oxon OX14 4RN

and by Routledge
605 Third Avenue, New York, NY 10158

Routledge is an imprint of the Taylor & Francis Group, an informa business

British Library Cataloguing-in-Publication Data
A catalogue record for this book is available from the British Library

ISBN: 978-1-032-32640-5 (hbk)
ISBN: 978-1-032-41826-1 (pbk)
ISBN: 978-1-003-35990-6 (ebk)

DOI: 10.4324/9781003359906

Typeset in Times New Roman
by SPi Technologies India Pvt Ltd (Straive)

Contents

Acknowledgments

Aquatopia is the result of a 12-year theatrical collaboration with Sofia Varino, probing climate questions, decolonial performative gestures, the nonhuman, and the historical past in search of an applied immersive methodology to engage with climate change as actors and directors. May Joseph thanks Sofia for the endurance, imagination, and optimism that fueled this slow manuscript. Huge thanks to the Senior Commissioning Editor of Routledge, India, Aakash Chakrabarty, whose vision, energy, support, and willingness to take a chance opened up this forum of Critical Climate Studies for the myriad non-academic approaches that accompany scholarly engagements addressing climate change. To Kavita Philip and A. J. James of the Critical Climate Studies Book Series from Routledge, India, immeasurable thanks for the rich and productive platform of their editorship to embark on this experimental writing journey about theater, landscape, and climate. Their supportive critique and suggestions made this manuscript a stronger book. Thank you to the rigorous outside reviewer for judicious suggestions that transformed the final draft. A special thank you to Tavia Nyong'o and Darren Suggs at the Park Avenue Armory for opening new possibilities for Harmattan Theater. Thank you also to Kimberly Jannerone of the David Geffen School of Drama at Yale for inviting me to engage with her theater students on climate performance. It takes our irreplaceable planet to make books, and this book in particular is the result of collaborations, generosities, friendships, and camaraderies of the extensive Harmattan Theater community that is spread across islands, coastlines, and rivers around the world. To all the friends, strangers, and family who joined Harmattan in the spirit of planetary concern from 2009 and counting, to create beautiful movement at Superfund sites and contaminated waterfronts, neglected shorelines, and forgotten navigational hubs, my deepest appreciation and thanks. We were performatively daylighting buried water sources before daylighting became a technique. Thank you, Harmattan Theater friends.

Theater is mostly tedious work, and I am indebted to many committed performers for their invaluable time in making Harmattan performances happen. While the community of performers exceed my ability to mention all names, special mention must be made of Sofia Varino, Celine Rogers, Victoria Marshall, Ariel Herrera, Jose De Jesus, Martin Gak, Alana Reuben, Lisabeth During, Patricia Hoffbauer, James Cascaito, the late David Van Leer, my brilliant documenter Tom Soper and photographer Joshua Kristal, Marista Miguel and Alipio Padilha in Portugal, Marit Bugge, Dil Hoda, Brian McGrath, Arthur Hutchinson, Russell Patrick Brown, Adam Lelyveld, Jennifer Telesca, Margaret Dessau, Khadeejah Gray, Dil Hoda, Ross Poole, Francis Bradley, Kristin Prevallet, Mathew Christian, and Nandini Sikand, for their joyous and public minded immersion in Harmattan's performances. To Diana Niepce, a special dedication for your extraordinary memorable performance at the Terreiro do Paco in Lisbon, forever in flight. To Harmattan's performers in Kerala and Cape Town, immense gratitude and awe. Finally, there is no theater without a reliable crew. My deepest thanks to my unshakeable crew, Geoffrey and Celine Rogers, for walking in rain and heat, dragging ektaras, mridangams, Tibetan gongs, food for the large cast, heavy props and umbrellas in hand carts, on subways, across the far-flung islands of New York City for over ten years.

May Joseph

Embarking on this 12-year project with May Joseph has been a creative adventure that *Aquatopia* as a book can only partly document. I thank May for all the fun we've had making it happen and for being such an energizing and uplifting collaborator. Many thanks to Routledge, most especially our editor Aakash Chakrabarty for his patience and dedication, and to all members and collaborators of Harmattan Theater (in particular guest performers Marit, Alana, and Diana) for playing with me in unlikely places at unlikely times. Thanks also to audiences and interlocutors at conferences where parts of this book were first presented, especially the Society for Science, Literature, and the Arts (SLSA), and the Association for the Study of Literature and Environment (ASLE). I'm grateful for the intellectual and artistic communities whose work has enriched and nurtured my own, and for family, friends, students, colleagues, and comrades spread across

New York, London, Lisbon, and Berlin—this book is also yours. I dedicate my aquatopian writing to Ramona for her canine companionship and sharp editorial instinct, and to Sam for their unwavering support and enthusiasm.

Sofia Varino

We also wish to thank all the performers, spectators, and sites we have worked with, in the past and in the coming future.

Earlier versions of portions of this book appeared in the following publications: M. Joseph and S. Varino. "Multidirectional Thalassology: Comparative Lagoon Ecologies," *Shima Journal*, Volume 15, Number 1, 2021; and M. Joseph and S. Varino. "Aquapelagic Assemblages: Performing Urban Ecology with Harmattan Theater," *Women's Studies Quarterly*, Special Issue: At Sea, Volume 45, Issues 1 & 2, Spring/Summer, 2017.

Preface

Aquatopia emerged as an ongoing public community conversation about how performance can productively engage with real world climate events to make sense of our dramatically morphing landscapes. Terrifying storms, flash floods, monster fires demanded that, as performers, dancers, actors, choreographers, we find ways to contribute to the ethical and environmental dialogue on how to live, heal, create, perform, and collaborate in this urgent time of climate awareness.

The recent United States Supreme Court Rulings of 2022, with their messages of anti-climate, anti-women's rights, anti-queer futures, and pro-gun legislation, remind us just how urgent it is that we persist in our small gestures of compassion, beauty, and creative collaboration to educate, cultivate, interrupt, debate, and transform society for the better, even when it feels dire and hopeless, as many of us are feeling in the wake of the Supreme Court decisions. The Clean Air Act and the Clean Water Act, the two great environmental achievements of the United States over the last 50 years, have been catastrophically compromised by the recent majority decisions of the US Supreme Court, with unimaginable consequences for global climate mitigation. *Aquatopia* is a small intervention in the conversation on why art is a critical tool for actively changing hearts and minds in the struggles to address climate change, and how we can continue to forge a shared future.

Prologues

Climate Precarity and Performance

By May Joseph

Humans are now in the throes of the climate gyre. We have to forge new ways to live under the threat of disrupted ecologies. This place of radical engagement is a space of performance, of provisional acceptance. For many who wish to act and do something about it, the enormity of the scenario of climate change has meant that the small act is more important now than ever. We are ethically bound to engage with the nonhuman—the sea, the forest, the desert. Humans, especially those in industrialized late capitalist societies, have systematically contributed to the precarity we find ourselves in at this point. Staring into the looking glass, we gasp aghast at how little we are ready to be accountable for, even as the hour for immediate action has already passed us by.

How Can We Engage the Anthropocene?

I began to grapple with this searching question in 2004 by creating performative engagements with oceanic sites following the devastating impact of the 2004 Christmas or Boxing Day tsunami. Monstrous storms had devastated societies before, but this earthquake catastrophe with its tentacular reach across from the Indian Ocean through the Bay of Bengal to the Malacca Straits was a deluge of monumental proportions. It was gargantuan in scale and seemed to draw together the Global South into a historical reckoning and a reconnecting unforeseen in ecological terms. Older colonial histories of port city logics were washed up onto the shores of saltwater-soaked regions, bringing with them a new era of coastal storm surge. Something irrevocably historic had just happened across the Southern Hemisphere—from Brazil and Mombasa to the Andamans and Aceh. It evoked a primal response of sorrow and horror. A call for action in some way. A physical embracing of the sea and its shorelines in a gesture of reaching out to the hydrosystem violently belched into the monsoon sphere of wind, storm, rain, and pour. An immersive submission into the vagaries of the planet begged a rational communal gesture, an act symbolic yet immediate and tactile, moving and transformative. Such an immersive submission would need to be simple, minimal, and accessible to the non-actor in order to be most effective as a vehicle of ecological movement. It needed a form that was closer to Bertolt Brecht's notion of process, of the work in progress, than the virtuosic product of Broadway entertainment. The coast from the Indian Ocean and the Bay of Bengal to the Malacca Straits was torn apart in 2004, in ways that the Global North was blissfully unaware of. For large populations from Indonesia and Bangladesh to India and East Africa, the sea had reared its ominous potentiality, one that rendered humans inconsequential and gods impotent. A new language of the theater needed to appear in the face of the new knowledge of the anthropocenic impact of history, and the anthropogenic accountability of modernity. This language would be as tentacular as its violent 2004 path. Many approaches, multiple strategies. But they would all have to begin to address the new language of ecology and oceans in some way.

Performance and the Anthropocene

The last decade has produced a vast amount of knowledge about how the Anthropocene and its impact on human futurity has opened up the chasm between understanding the nonhuman and ethically engaged social interventions. Scholars, scientists, and environmentalists have made us aware that one of the challenges facing our understanding of oceans and the planet is the anthropocentric nature of our engagements. The planet has been transformed by human actions into an extractive zone without accountability. We are now staring at the widespread destruction of human and nonhuman life exacerbated by unfettered capitalist overdevelopment and the consequences of centuries of colonial violence. A plethora of scientific studies confirm that all forms of planetary recovery are needed, from anti-consumption to carbon mitigation to climate adaptation, in order for life as we know it to survive its own self-destructive impulses.

Performance offers itself as a methodological, political, and ethical intervention in the face of searing uncertainty. Drawing on performance as a sphere of ecological practice, of engagement, of immersion, to reconnect the earth to human sociality in its more responsible iterations is one technique of anthropogenic pedagogy. What are the ways through which performance and its processual acts of performativity throw open opportunities for investigating the junctures of embodiment, environmental ethics, and the Anthropocene?

Planetary potentiality is the miasma of performance. The proscenium has opened outward, embracing its historic oceanic vistas as in the theaters of Dionysus, of Delphi, and of Ephesus. Oceanic preoccupations have entered the *mise en scène*. Water has deluged the planet furiously in the twenty-first century, and its liquid traces have foregrounded a new historiography of hydro futures. Where the ocean was the distant horizon during the Greek period of dramatic imagining, its fury has swallowed up the land between ancient waterlines and contemporary landfills, leaving new lakes, seas, and bays in its wake.

Performance as a lens for climate change emerges as an organic phase of climate adaptation, where societies unprepared for sea-bound calamities and the morphing ocean, are provisionally, temporarily, performatively inventing ways to address the new reality of rising seas. This experimental performative aspect of climate actions foregrounds performative acts and performance as methodology in new contexts, harnessing older histories. Performance and performativity are central to a pedagogy of the Anthropocene. The highly contested concept of the Anthropocene, as proposed by climate scientists, refers to the period of escalating industrialization that from the sixteenth century onward systematically transformed, destroyed, and overdeveloped vast regions of the earth in ways that went unaccounted and undervalued.[1] Five hundred years of planetary overuse has brought the earth's diverse species to a series of multiple extinctions. At this juncture, the versatility, the interactivity, and the sociality of performance presents itself as a tool of hope and constructive imaginaries even as the assessments of gloom and dire consequences muddy the ontological search for a new language of performance in the Anthropocene.

Precarity and the Theater of Climate

We are living in the era of provisionalities—of making plans, arrangements, feeble adaptive gestures towards a culture of contending with the new normal: too much wind, too little snow, too much rain, too little water, too much heat, too much cold, *too much weather*. It is the age of a new normalizing global anxiety in which performance occupies a new ethical possibility – that of engaging, intervening, and responding in the age of uncertainty. The theater of precarity has transformed how we think about gesture, action, engagement, and performance into a new sphere of urgency. It is critical that as citizens engaged in a world of ecological injustices and environmental degradation, we ask ourselves how we can intervene, impact, alter, or change ecological understanding wherever we are. This precarity of a shared human future foregrounds the role of action, enactment, engagement, and activism on new terrain. Societies living along vulnerable coastlines are involved in engaging with the new realities of a changing volatile climate. The need to respond through performative acts in public spaces, whether through marches, speeches, performances, encounters, meetings, or digital hangouts, has expanded the implications of performance for climate change. Performance is a tool for education, for outreach, for awareness building, as much as it is the practice of the lyrical.

What drives the need for performance is the raw force of nature itself. We are now in a terrifying place of recurring climate anomalies. Precarity and its impending effects of fear, uncertainty, unknowability, are the hidden economy of the contemporary moment. Populations across the Global South live with this creeping dissonance of a climate out of kilter, a predictability increasingly thrown into volatility. Accompanying this precarity is the attending set of active measures, of building seawalls, of bolstering levees, of reinforcing eroding buffer zones, that creates a vernacular of climate adaptation across low lying coastlines. We are suddenly immersed in a sinking world whose depths we took for granted. Our ecological hubris is our motivation for the new theater of precarity. It is the space of interrupted, ephemeral, nervous enactments – architectural, social, visual, sonic, visceral – through which our societies are beginning to measure up to the enormity of inundation. Performing in this space of submergence is to fight to stay alive, to overcome our myopias to envision a livable future.

Inventing an Oceanic Practice

In 2009, I created a theater company in New York City to produce engagements with coastlines and precarious landscapes. This urge, precipitated by the 2004 tsunami, is the gesture of Gaia, the immersion with earth's potentialities.[2] Creating theater to engage with climate volatility is one way to respond to the emerging scenarios of impossible weather taking its toll on landscapes that have been tamed by humans. The act of theater, however, is an ensemble gesture. This, in the age of capitalist productivity, hinges on the largesse of other utopians who believe in the positive impact of non-utilitarian actions on the planet. Such a gathering of individuals coming together to generate a performative solidarity of care and responsibility for the planet in real time is one of the most powerful sensations one can experience in the face of conflicted hopes of mitigation. We must rely on our senses and on our bodies to keep asking questions of nature and of the nonhuman around us through movement and choreography. We are immersed in their vitality, and embedding ourselves in their anima offers possibilities for collective aid and endurance that we are fast forgetting exist. Creating a theater company was a way of producing a conduit to engage with climate change on its terms, to ask questions of landscapes and open waters that were not possible within the confines of the stage itself. The ocean needed to be drawn into the sphere of the human, and to be interacted with performatively. An oceanic praxis needs to be reinvented.

An oceanic practice entails a listening and submission to the sea and its depths. It is a cultivation of a new relationship with the sea that is still bound by the limits of the shoreline, of the coast, made attentive to the shifting sands and receding waters of daily life. An oceanic practice is a quotidian practice. It is about the constant awareness of the ocean's relationship with our lives beyond its extractive peripheral existence as a terrain of transport, source of sustenance, a sphere for taking without account. An oceanic practice is a symbiotic engagement with the nonhuman ontologies of becoming that shape coastal, water-bound communities, and the ways they navigate their landscapes. The oceanic is a mode of being, of becoming.[3] It is both internalizing the histories of the sea's pasts, its legacies and hidden consequences, as well as embracing its present turmoil and volatility. Our societies need to become more oceanic. To be alert to the ecological, historical, indigenous, and social justice implications of the ocean's interface with human life. This is particularly striking when one visits cities like Honolulu, whose island-ness has not generated a sense of the oceanic. Honolulu has been disconnected from its oceanic sensibilities. Its present status as an extractive zone whose surroundings – the coast, the sand, the forest, the sea – are entirely geared towards rapacious tourism, marks the difference between a city *by the sea* as opposed to a city *in conversation* with its oceanic ecologies.

Honolulu is the poster child for the disengagement with the oceanic. The area surrounding the city of Honolulu has been mindlessly built over, its water sources buried under macadam, its wetlands, river flows, and oceanic contours turned to concrete in its surge towards hyperdevelopment. Now it faces the vulnerabilities that come from not attending to the island's geologies regarding water management and water resources. The aquifers around Honolulu have been drained by developers leaving local farmers without water for their agricultural needs. In an island rich in natural resources, a condition of water scarcity has arisen because of unsustainable colonial exploitation whose implications for future land use are creating an explosive political impasse between indigenous anti-colonialists and mainland settler colonialist developers.[4] Visiting Honolulu clarified for me what an oceanic thinking is not. Where the ocean has been silenced temporarily with military grade hardscape. Standing at the edge of the sea in Honolulu, one is far from the oceanic. The eroding beaches, the fake sandy terrains of Waikiki speak of an ailing oceanic knowledge that needs to be listened to and recuperated once again. This wail of the oceanic is very strong in Honolulu as one stands surrounded by concrete, metal, and single-use plastic floating across the island's commercial establishments. An oceanic praxis is a multipronged involvement with the totality of coastal living, at once personal, political, urban, local, performative.[5]

Opening the World—Climate for Real

By Sofia Varino

A decade is something to think about, something to think with, a chunk of time substantial enough for narrative arcs and patterns to emerge, but not so vast that it becomes impossible to grasp and account for. Writing this book demanded a continuous return to performances past as well as creative engagement with the pandemic present, which we have all endured under such disparate conditions and degrees of privilege. May and I started working on the book that has become *Aquatopia* shortly after our performance and joint talk at a conference in Venice was canceled due to COVID-19, in April 2020. During the long months of writing from early 2020 to the end of 2021, we reviewed dozens of Harmattan performances, often with longing for times gone by. But perhaps the writing really began as we assembled and reviewed our rehearsal and preparation materials alongside media documentation of the many performances we have collaborated on since 2010: notes, diagrams, travel plans, shared itineraries and schedules in parallel with Harmattan's multimedia archive encompassing video, photography, and text.

When a book is the culmination of such a long and sinuous process, it becomes charged with expectation and ultimately the recognition that there is only so much that the medium of written language can convey. The limits of language, what it cannot encompass or reach, are partly why I first began using performance in 2003 as a strategy for articulating collective and personal experiences I could not find words for, from navigating queer and sex-positive subcultures from the late 1990s to my encounters with the wilderness of the Atlantic Ocean or the breathtakingly raw beaches of Loføten in Northern Norway. There were older memories too, from overflowing industrial pollution of the Tagus River during flood season, the vast estuaries and vineyards, the windmills and the fruit trees, the musk of grapes rotting into late summer, which had stayed with me from my childhood in post-revolution 1980s and mid-1990s Portugal. From 2003 onwards, performance quickly became a social experiment and an outlet for whatever was unbearable in the London queer underground I was part of: gentrification, rising rents and rampant poverty, squats and illegal clubs, sex work, building chosen families and relationships against heteronormative expectations around monogamy and child-rearing, rape and sexual violence, pleasures that made their way underground in burlesque, strip tease, sex parties, and BDSM events.

Pandemics and Uprisings: Socially Distant Kinship

The first Harmattan performance I participated in was *Dreamscapes* in New York in 2010, with the companionship of the Hudson River, the Chelsea Piers, and the many life forms that inhabit or traverse the site. The ensemble choreography, with a line of upright human figures dressed in white against a backdrop of river and sky, the slow speed of the movement of our bodies pushing through the landscape, delicate shapes suspended in time, was documented in a series of black-and-white photographs by Joshua Kristal. By 2010, new calamities had been making their way into queer and feminist circles and were permeating modes of queer resistance, adding to and expanding the work of AIDS activism as retroviral therapies gradually became the norm in wealthy Western nations. Concerns for climate change and environmental devastation were of course not new to queers and certainly not to feminists, nor were connections to the important function of public spaces and urban life as sites of political action, community organizing, rampant celebration, and collective healing. The financial crisis of 2009 had made all these concerns more urgent and tangible in the USA, both fueling and hindering anti-capitalist modes of resistance, weakening most people's capacity to endure, while fostering a wave of social and activist creativity, eventually culminating in the Wall Street Movement.

I am borrowing the prescient phrase "from the realm of pandemics & uprisings" from adrienne maree brown's foreword to *Undrowned: Black Feminist Lessons from Marine Mammals* (2020) by Alexis Pauline Gumbs. That is the realm indeed which we have been inhabiting and learning to navigate, one demanding physical distance and containment, the other requesting our collective social energies in coming together to denounce, resist, transform, and transgress the deadly machinery of colonial heteropatriarchy under late capitalism. We, those of us alive on this planet, right here and right now, desperately need an in-depth political reconfiguration that can effectively counter and disable the lethal patterns of extractivism and exploitation, while repairing its devastating effects on so many lives.

The unprecedented fluvial floods in the western region of Germany during the summer of 2021 are a sign of how climate change and environmental crises will not only impact sea-bound inhabitants. Wealthy European nations hoping to escape the consequences of environmental devastation due to geographical location or economic prowess must now grapple with the fact that this is not the case. And at the time of writing, the German Green Party (die Grüne) has been elected to be in a three-party coalition governing one of Europe's wealthiest nations.

The Chelsea Piers encapsulate their own living archive of downtown New York City (NYC) queer histories and perform their own daily labor as iconic sites of Indigenous history prior to the colonial period. The island of Manhattan today is part social lab, part architectural experiment, and part living museum. Environmental devastation and slow violence (Nixon) mingle freely with spectacular sunsets, unforgiving hurricanes and snowstorms, blistering hot summers. The island is theatrical in itself, re-enacting in its water-boundedness the aquatic condition of a planet with two melting poles and rising sea levels. For me, the Chelsea Piers had strong associations with my first NYC Pride celebrations in 2006, with its by then corporate, gentrified, and heavily commodified scenes interweaving cruising, BDSM, clubbing, and drug subcultures. Historically, NYC stood for me as an icon of postwar immigration, a monumental archive of the consequences of warfare and genocide, of persecution and intolerance, and later of greed and exploitation. To cross that slice of land protruding into the water, walking into the sunset dressed in white, struck me as emblematic of NYC's divergent genealogies, of its accumulated indigenous histories, the multiple violences of its subsequent European colonization, its iconic status as an economic and cultural hub, and its subcultural heartbeat. As I look back now, that experience encapsulates for me how performing queer-belonging beyond citizenship can also be a decolonial method: by redistributing the geological, animate, and technical resources of a space, harnessing its powers away from a heterosexist patriarchal capitalist system of (re)production and converting these resources into a vibrant assembly energized by social relations, structures, networks. The social energies Harmattan harnesses and generates in collaboration with sites and audience form the basis for the company's decolonial queering of space as performance praxis.

In deep solidarity with direct action and protests, Harmattan as a troupe has sought to offer alternative venues for environmental engagement centered on historical accountability for the past that haunts the sites we work in; aesthetic attunement with the present moment and the presence of the elements, of weather and life forms, and the built environment; and a creative politics oriented towards imagining possible, though not certain, collaborative multispecies futures. Since as performers we are also involved in doing the political and intellectual labor of protesting, writing, documenting, teaching, and resisting climate change, the work that Harmattan presents is more an extension of than a departure from more direct environmental protest interventions.

The Future of Water

Bound to modern science and its technoscientific paradigm, there is much that current environmental activism, science, and philosophy cannot yet articulate. Similarly, environmental movements remain limited to party politics and the model of a representative democracy that is anything but democratic. Citizenship is determined by the military and economic interests of the nation-state, and industrial and commercial profits determine the wealth of nations, of populations, and of budgets for mitigating climate change. Some of these measures are clearly beneficial, if they are vigorously implemented and can benefit the most affected populations, rather than creating climate, food, and water security for the citizens of wealthy nations. At such a crossroads, why turn to performance? Or, to ask it another way: what is so irresistible about performing that we turn to it in such desperate times, *in spite of everything*? Engaging with Brazilian philosopher Denise Ferreira da Silva's concept-practice of "poethics," which draws on how the Portuguese word for poetics, *poética*, contains the Portuguese word for ethics, *ética*, Harmattan creates performances addressing the earth archives of harm inflicted on human and nonhuman communities.[6]

These earth archives are quite specific ones: as harm gets sedimented in the geological remains of centuries of genocide, colonial violence, environmental damage, and the harm inflicted by the extractive logic that turns all things into their surplus value: earth into land and life into labor.

A Very Brief Chronology

My own brief chronology: in 2010, I joined Harmattan Theater at the Chelsea Pier to move along with others with the afternoon riverside commotion of NYC life, a signature blend of protest, community organizing, public assembly, leisure, art, and urban activity. In 2013, I become Associate Director. I write this prologue in Berlin as a permanent resident and soon to be national of Germany, as an EU citizen, as a Portuguese-born queer Jew who's spent most of their life living abroad: after Lisbon, London, Sydney, Perth, Bergen, and New York were all cities I called home. When I joined Harmattan in 2010, I was living in New York quite impermanently on a student visa. I'd just started my graduate studies at SUNY Stony Brook and was working as a teaching assistant for a BA course on Literature and Sexuality. I'd already been living in New York since 2006, barely managing to hold on to its razzle dazzle, performing downtown and working odd jobs, editing and translating while completing an MA in Theater at Hunter College. I met May when she gave a guest lecture on performing ecology at the inimitable Una Chaudhuri's Performing Beyond the Human MA seminar. Some courses are life changing. I had already completed all the credits necessary for graduating from the program, but the Beyond the Human in the course title caught my eye. I'm glad it did, or else I would not be writing these words now. Something May said stood out for me: she talked about the particular obstacles to creating ensemble work that is more visually oriented conceptual performance art than text-based psychological drama. Performance artists do solo work, or perhaps collaborate sporadically with other artists. But the joys and challenges of systematic, collectively devised work are rarely available to performance artists. I longed to do ensemble work that was conceptual and movement-based in public spaces, accessible to performers of all levels of skill and ability, and to spectators of all levels of income. That this work should be politically and ethically invested in decolonizing spaces while raising awareness about climate change made perfect sense to me. And that our aesthetic arsenal included Augusto Boal alongside Fernando Pessoa, performance theory alongside traditional Indian theater techniques, was a boon.

The relatively comfortable existence of so many privileged citizens living in Western and industrialized affluent nations is only possible due to the energetic and material depletion of people and resources in other regions. And amidst the extremely unlikely conditions for life on earth to emerge, it is a positively odd occurrence that elaborate and extremely delicate life forms, such as mammalian, reptilian, and amphibian vertebrates, can nonetheless persevere and even thrive. Harmattan harnesses these paradoxes to shape climate performances that seek to unsettle rather than resolve the current paradigm of planetary uncertainty instigated by environmental devastation. Harmattan's queer decolonial praxis involves living with and enlivening sites whose status ranges from abandoned, forgotten, vilified, to iconic, at once toxic and scenic. These sites demand and foster intimate modes of relationality: an intensified proximity through their condensed histories, the haunting of memories and mementoes they contain, further activated through our collective slow meditative walk, facilitating immersion and hyper-sensorial attunement to experiencing in a heightened manner a familiar everyday space bustling with urban life, an iconic historical site swarming with tourists, or an abandoned dilapidated pier left to its own decay. So, the question remains: why perform at all? What can performance do as climate intervention? Finding ways to keep asking these questions is what the following pages seek to do.

Notes

1 For a foundational discussion of the term "Anthropocene," see Paul J. Crutzen, "Geology of Mankind," *Nature* 415 (2002): 23.
2 See Bruno Latour, *Facing Gaia: Eight Lectures on the New Climatic Regime* (2015), Cambridge and Medford, MA: Polity Press. See also Stephen Clark, "Gaia and the Forms of Life" in *Environmental Philosophy*, edited by Robert Elliot and Arran Gare (1983), University Park, PA: Pennsylvania State University Press.
3 See K.A. Ingersoll, *Waves of Knowing: A Seascape Epistemology* (2016), Durham, NC: Duke University Press.
4 See Macarena Gomez-Barris and May Joseph, "Coloniality and Islands," *Shima* 13:2 (2019): 1–10.
5 See Hŏkūlani Aikau and Vernadette Gonzalez, "Curating a Decolonial Guide: The Detours Project," *Shima* 13:2 (2019): 11–21.
6 See Denise Ferreira da Silva, "Toward a Black Feminist Poethics: The Quest(ion) of Blackness Toward the End of the World," *The Black Scholar*, 44:2 (Summer 2014): 81–97.

1 Storm as Method

Climate Performatives

May Joseph and Sofia Varino

Harmattan Theater is a multidisciplinary New York City based environmental performance collective. Harmattan Theater's experiments since 2009 have been a submergence into the waves of subsidence, of deluge, of waterlogged consequentiality. Founded in 2009 by May Joseph, Harmattan consists of a voluntary group of dancers, actors, composers, musicians, architects, designers, writers, scientists, environmentalists, philosophers, and artists committed to site-specific performances addressing climate change and the future of water. The collective has devised 21 movement pieces presented at various precarious water-bound sites affected by rising sea levels. Harmattan Theater has to date presented 18 large scale, site-specific projects in places like Istanbul, Cochin, New York, the Maldives, Venice, Rome, Lisbon, Amsterdam, and Cape Town, connecting these water-bound locations through their maritime histories, migration flows, and geographical vulnerability to rising sea levels and climate change. Collectively considered, the projects form an ethnographic and historical excavation of fort and port cities whose histories shaped an extensive part of the early modern period from the eleventh century on and whose future is impacted by a new global threat: human-made environmental degradation.

Climate Ethnography: Opening the World

How do pedagogy and performativity come together and inform one another in an oceanic praxis? How can performance methods be deployed in pedagogical contexts to support environmental attunement, from formal education (beginning with kindergarten, or perhaps even earlier, to higher education) to public events, community organizing, protests, and ephemeral social gatherings? The kinds of terrestrial performatives Harmattan proposes are of the earthly,

DOI: 10.4324/9781003359906-1

urban, gritty and sweaty, sticky messy kind. We oppose the smooth slickness of the neoliberal greenwashing paradigm as much as the purity of ecological conservationist dogma. Nothing needs to be preserved or protected as just another kind of commodity or resource, something that must have value to have worth. Rather, non/human rights should be implemented. Each human life is an ecology embedded into many other interconnected ecologies. To decolonize the ecosystem of a living body means paying attention to how historical forces have either depleted or supported its aliveness. If ours is a queer decolonial climate ethnography performed in real time, it is an ecology of the wild, the feral, the unbound. Earth dwellers need tangible solutions for mitigating the effects of environmental destruction—and we need them fast. But we also need imaginative dislocations, tearing apart what appears as given and unchangeable, but which is in fact the result of a long chain of historical events, random and contingent.

"Aquapelagic assemblage" is a term proposed by Phillip Hayward (2012) to identify processes of transformation along aquatic ecologies at the interface of human sociality and marine environments. The performance work of Harmattan Theater has involved a persistent inquiry into the junctures between colonial history, coastal landscapes, island ecologies, and human sociality. Drawing on the idea of "aquapelagic assemblage" as an aesthetic practice, in this chapter we explore Harmattan performances since 2009 as historical experiments towards a praxis of such assemblage. According to Phillip Hayward (2012), aquapelagic assemblages are water-bound interfaces between animate and inanimate entities. They occur performatively at the threshold between social life and environmental landscapes involving aquatic spaces. These liquid assemblages are initiated by human engagement with island topologies in specific historical relations. Harmattan Theater's performances are an oceanic praxis. The ocean, the coast, coastal communities, and climate adaptation are the roots of how this theater as event has come about. Harmattan's immersion in the practice of creating aquapelagic assemblages along historic maritime locations of the colonial era is generated with the purpose of reassessing the past while interrogating the future. Through performative encounters with maritime history, Harmattan's assemblages open up a chain of geospatial continuities across the planet.

In Harmattan's choreographic explorations, the world is a network of oceanic entanglements. Pursuing the symbiosis between cultural memory and the ecology of island geologies, its movement work is motivated by climatological inquiries such as: How has water impacted modern city making? How is the proximity to the sea transforming coastal communities today? How is climate adaptation redesigning medieval port cities? What are the ecological relationships between the colonial port cities of the Dutch and Portuguese era of the fifteenth century and contemporary postcolonial cities? Performance is a powerful tool to deploy in sites of *acqua alta*, high water. As theater practitioners work with islands and archipelagoes, we have learned from performing in Venice that the world is increasingly becoming like Venice in its inundation experiences. Addressing high water (or, as the Venetians call it, *acqua alta*) as a global hazard opens up the comparative understandings of how to live with flooding, sinking, and oceanic deluge. The work of Harmattan Theater with coastal and island populations has exposed the versatility of the language of performance to connect languages of science with vernaculars of local ecologies. Performance enables diverse publics to engage with their environments in unexpected ways, allowing them to ask questions, participate in their own landscape, interact with their coastal environs, and enjoy their precarious seascapes. For Harmattan, these are interactive journeys of historic reclamation that recover the past with questions impinging upon our urban futures.

If modernity emerged at the cusp of seafaring histories and land-bound migrations, then the contemporary era of the Anthropocene (a hotly debated and highly contested term proposed to designate a new current geological era marked by the impact of human activity on the planet) suggests a web of interconnectedness among human and non-human forces vaster than previously thought. Nonetheless, the term pragmatically informs Harmattan's work in multiple ways, especially as a key term for signaling a condensation and intensification of interconnected global phenomena since the beginning of Western colonialism until today. For Harmattan as an ensemble, performing in, during, against the perils of the Anthropocene gives our work a register of urgency and haste. Our rehearsals are rushed, our preparation rudimentary, our shows devised in a hurry, performed once or twice and documented in real time as video and photography. Before you know it, our traces have vanished from any site we have worked with for a few days or a few hours. The performers leave and spaces return intact to being whatever they were already. We perform in a constant state of emergency and vigilance. We arrive, perform, and vanish.

First coined by American biologist Eugene F. Stoermer in the early 1980s to denote the effects of human activity on a planetary scale, the term "Anthropocene" was later deployed and popularized by Dutch meteorologist and atmospheric chemist Paul J. Crutzen, who introduced it into mainstream scientific discourse in 2000 to designate a new geological epoch characterized by the widespread effects of human activity on the atmosphere. Taken up by scholars in the humanities and social sciences, the term has seemingly acquired a life of its own, signaling everything from climate change to rapid species extinction, deforestation, and habitat loss. Its imprecision is precisely what makes it so appealing—after all, we are living in unprecedented times, for which we barely have adequate language to analyze data and describe phenomena. Perhaps most importantly, the imprecision of the term allows it to encompass that most elusive category of variables: what we do not know we do not know. Whereas the term "multispecies" shares a common genealogy around environmental concerns and ecological thinking, it does not (yet) carry the same degree of scientific authority or cultural capital.[1] Whereas a multispecies approach aims to decenter the human and highlight the multispecies entanglement of all life forms, the Anthropocene does not force us to confront how underlying Western epistemes have contributed to the problems facing earth dwellers, nor to (re)imagine how to counter its pervasiveness. Although it may signal some of the overall problems, the Anthropocene does not invite us to image a world otherwise. As a term, it manages to contain a vast array of the symptoms plaguing ecosystems today, but the Anthropocene as a concept fails to function with the precision of a diagnostic term like "climate change." The Anthropocene refers to the deadly effects that the awe-inspiring discoveries and inventions of the United States and Europe have had worldwide. In 2000, Crutzen and Stoermer proposed the latter part of the eighteenth century as a potentially adequate date for marking the onset of the "Anthropocene" although they acknowledged the date as somewhat arbitrary and that alternative dates could also be considered, even up to the beginning of the Holocene epoch:

> However, we choose this date because, during the past two centuries, *the global effects of human activities have become clearly noticeable*. This is the period when data retrieved from glacial ice cores show the beginning of a growth in the atmospheric concentrations of several "greenhouse gases", in particular CO_2 and CH_4. Such a starting date also coincides with James Watt's invention of the steam engine in 1784.[2]
>
> (Emphasis added)

Although the term has yet to be officially recognized, on August 29, 2016, the Working Group on the Anthropocene (WGA) recommended to the International Geological Congress the formal designation of the current epoch as the Anthropocene. As an era marking the repercussions of the human presence on earth, particularly nuclear bomb testing and plastic pollution, the Anthropocene moment is a contradictory one, in which former colonialisms and contemporary globalization merge into a larger historic trajectory of environmental exploitation and ecological impact. How human activity has radically transformed the planet is deeply embedded in globalizing colonial histories, amply evident in the ruins and urbanization of water-bound cities. For example, the ancient and medieval history of Venetian, Portuguese, and Dutch trade and colonization between the twelfth and seventeenth centuries has left its traces in over 600 colonial entrepôts around the southern hemisphere, and these contact zones in turn impacted the metropolitan centers of Europe.

In *Fluid New York* (2013), Joseph traces precisely these complex aquatic topologies in the New York City landscape, whereby the colonial past becomes entangled with the contemporary everyday assemblages of urban life in a web of vibrant collective activity. Forgotten Lenape routes are transformed into major Manhattan roadways, for instance. Erased relationships to the wetlands of New York are slowly being revived through a renaissance in oyster beds and mudflats around New York's shoreline. Joseph's arguments in *Fluid New York*, that the city of New York entirely ignored its precolonial and colonial archipelagic topologies until the onslaught of Hurricane Sandy in 2012, document the shift within American urbanism towards a more coastal driven approach to waterfront development along the eastern seaboard of the United States. Joseph points out how a car-bound Robert Moses's twentieth-century approach to urbanism has morphed towards a more environmentally attenuated one over the last decade. For Harmattan Theater, this slow embrace of New York's water-bound topographies have been the grist for playwrighting, choreography, and performative activisms. We have sought to animate the buried and hidden water sources of New York's historic waterlines through site specific performances of duration. Slow walks through forgotten coastal landscapes.

At the end of her influential *Vibrant Matter: A Political Ecology of Things* (2010), a treatise on the political agency of nonhuman entities, Jane Bennett proposes a "credo" for vital materialists, summarizing the central tenets of an understanding of materiality as lively and enlivening, and the ethical implications of such a position: "I believe that encounters with lively matter can chasten my fantasies of human mastery, highlight the common materiality of all that is, expose a wider distribution of agency, and reshape the self and its interests"

(122). In line with Bennett's vitalist credo, Harmattan Theater's projects enact the premise that water is a substance whose political agency participates in the public sphere, shaping the everyday lives and the futures of humans and nonhumans alike. By paying attention to natural and unnatural histories, Harmattan seeks to explore new strategies for ethically engaging with water cycles and sites, for finding new cultural and aesthetic forms that awaken and deepen our understanding of our environments, and to begin to envision sustainable urban futures.

Harmattan's assemblage works are historical engagements merging maritime navigational history with climatology and marine ecology. In this book, we document and analyze how Harmattan's aquapelagic aesthetics enliven and engage with the already highly active, performative environments of water-bound cities. Pursuing the connections between rising ocean temperatures, colonial pasts, and the reinvention of port cities today, in 2009 Harmattan Theater began a series of site-specific performative installations around former Italian, Portuguese, and Dutch colonial ports whose histories shaped the Malabar coast of India, where Joseph grew up, as well as the New York City cityscape where she currently resides and Harmattan Theater is based. Members of Joseph's family, originally from Cochin, India, via East Africa, have Portuguese last names, a final vestige of origins, trajectories, and becomings.

Performing Port City Ecologies: Dutch and Portuguese Legacies

One of the most persistent responses to Harmattan's projects has been that perhaps they recite, without sufficient critical distance, the very archives of colonial exploitation, brutal capitalism, and ecological devastation that we aim to critique. Although our performances do make explicit reference to these violent archives, and are at least partly catalyzed by them, we believe we must confront the actuality of the histories we encounter, their pervasive presence in our everyday lives—materialized in architecture, geography, urban infrastructure—and the disastrous effects they still carry for vulnerable populations around the globe. Sites carry historical weight; they are memorials and containers, receptacles of the collective experiences that shape the stone and brick, gravel and grass, salt and concrete, air and water, extended over time and space. For example, Cais das Colunas in Lisbon is as emblematic of exploitative sea economies during the fifteenth and sixteenth centuries, a materialization of Lisbon's colonial past, as it constitutes a meeting point for random publics and improvised communities, a site of leisure, tourism, and entertainment, of artistic activity and social exchange. Street culture emerges and thrives by the riverside, mixing the city's colonial pasts with the immediacy of environmental decay, toxicity, and industrial pollution; the precarity of exposure to rising water levels is mixed with urban renovation, construction, experimentation, and creativity. When the sea sprite of *Mar Português* and the ghost of Vasco da Gama's daughter move across this haunted pier, they meet and let each other go, tracing a trajectory in trance, marking with their footsteps a larger maritime narrative of seafaring, global trade, industrial waste, and volatile transnational encounters.

Since oceans and bodies of water constitute the central theme of the work we do, the precarious state of these ecological sites, their vulnerability and exposure to the natural elements, becomes a key part of what the audience and the performers are invited to experience and enact, live and relive. The future of water ceases, perhaps only temporarily, to be an abstract notion, as one faces harsh winds and tastes sea salt, inhales toxic fumes and shivers from the cold, breaks into a sweat or is blinded by the low sun over the course of a Harmattan performance. But perhaps more importantly, the ludic component of our work, the playful invitation to interact again with the elements, to draw figures with water, to move one's body in new patterns through familiar and unfamiliar spaces, to slow down and walk with intention and conscious awareness—all these offer opportunities for encountering urban seaside and riverside areas in new ways, to reconsider our reliance on, oblivion to, or total avoidance of these exposed areas. These sites are not only bearers of the heavy, at times nearly unbearable, weight of their histories, nor can they be simply dismissed as concentrated points of economic value and cultural capital. And they can never be reduced to an impending destiny of devastation, storm and surge, progressively swallowed by a high tide. Alongside their convoluted pasts and threatened futures, they remain, at least for the moment, viable zones of leisure and entertainment, communal areas of respite, sociality, and creativity.

Decolonial and Queer Performatives

These past 12 years of writing, reading, and performing about climate change have been informed by the methods, theories, and politics of decolonial and queer interventions across academic, activist, and artistic milieus. Our critical engagement with the terms "decolonial" and "queer," both over-exposed and over-used terms (to the point that it could be argued they barely mean anything anymore) has been less about the epistemological possibilities they present, although both have been shaped by rigorous scholarship and invigorating political demands across areas like postcolonial theory, critical race studies, environmental justice, ecofeminism, affect theory, and queer ecology. The work we have been developing with Harmattan has been no doubt shaped by all these, but our decolonial and queer commitments have been primarily ontological—What might a queer decolonial ontology of climate, of history, of place, of embodied life, entail? How can the primal, elemental forces of the ontological be harnessed into ethical engagements with large-scale phenomena that defy comprehension on a human scale? And yet, they encapsulate long genealogies of critical and political gestures we want to galvanize and, perhaps most importantly, acknowledge in how they have shaped our work as a performance ensemble. However, both terms have also been deployed as radical political interventions demanding a vast reconfiguration of power structures and systems of knowledge, especially in the early stages of their use in contexts of emancipatory politics.

"Decolonizing" refers to ongoing anti-colonial and anti-racist struggles that continue to demand a reconfiguration of non/human networks towards less deadly and damaging interactions. "Queer" has been on the one hand restricted in its use to identity politics, and a viciously exclusionary kind of identity politics in many cases. On the other hand, it has been so diluted across "high theory" critical contexts to the point of meaning nearly nothing at all, or at least nothing precise or definable, as a critical term/gesture should hopefully be, at least to some extent. And so we are using "queer" here to refer to the verb, to queer/queering, rather than the noun, as a way to signal towards the fluidity and elasticity of queering as an ongoing political project and critical practice that unfolds and spills over, supporting and shaping our work in unexpected, transient ways.

Although Harmattan's methodology is broadly situated within practices like experimental ethnography, oral history, community theater, and practice as research, we remain committed to relational, participatory, flexible, and most importantly, decolonial performative structures. Embedded in the performative, participatory structures of conceptual art and relational aesthetics (a term proposed by Nicolas Bourriaud (2002) in his influential work *Relational Aesthetics* to designate artistic practices that emphasize social structures and relations), Harmattan's performances and videos can be placed in turns within the context of decolonial environmental theater, immersive art, and situationism. Theoretically and thematically, our work is informed by political ecology, postcolonial studies, decolonial theory, and experimental geography, alongside New Materialisms and the critical turn of postmodern/posthuman engagements with the lived environment as proposed by works like Baz Kershaw's *Theatre Ecology* (2007), Jane Bennett's *Vibrant Matter* (2010), or Andreas Weber's *Enlivenment* (2013).

Our projects are investigations into oceanic histories through real-time encounters with urban ecosystems, excavating the cultural weight of the landscape in its intertwining with global postcolonial marine space. Water-bound cities and shorelines, seawalls, waterfronts, barrier islands, and dilapidated piers form the crux of Harmattan's encounters with liminal coastal urban areas, presented both in live performances and videos. The Harmattan multimedia database encompasses a broad range of written and visual materials, including centuries-old archival documents like maps and treatises, alongside visual archives of performances and interactions with local human and nonhuman inhabitants. Retracing maritime memories and diasporic trajectories, the Harmattan ensemble enacts a flow between fragments, dreams, maps, documents, letters, interviews, and performances as public repositories of the past.

In the vein of art activism and participatory aesthetics, each Harmattan project is devised as an interactive event in collaboration with local musicians, dancers, poets, tourists, passers-by, and residents who participate in the performance after a brief training session in the slow walking meditation techniques of "Vipassana," a Buddhist practice popularized by the Vietnamese monk Thich Nhat Hanh. Influenced by the forest walking traditions of Laotian monks with whom Joseph studied in Laos, Harmattan introduces the participant into the method of the slow walk, the being in the walk of the Vipassana technique. Harmattan devises partly improvised, yet highly structured, urban spectacles where the cosmopolitanism and modernity of performance art and experimental theater are combined with traditional performance styles like Indian and Tanzanian environmental and folk theater traditions, including Jatra, Chautu Nadagam, and Indian street theater. Throughout the preliminary stages of research in preparing a project, we pay careful attention to historicity and place. The various archives we excavate through our research, which are later to be reenacted live, are often volatile ones, materializing the thick, convoluted histories of sites like Lisbon's Terreiro do Paço, where enslaved people were traded alongside merchandise, or sinking Venice, lingering on the verge of disappearance with each seasonal flood.

Our live encounters with these sites become permeated with the stickiness of the past we aim to critique and must nonetheless revisit. Performing with Harmattan Theater becomes a collective work of excavation, of extracting both memories and fantasies from the various strata of each urban site: buildings and built environments, alongside mineral and organic life, local climate and natural elements. The agency of water, the elements, and organic and inorganic matter become part of a performative assemblage, as wind, waves, sun, rain, earth, sand, and stone all play decisive roles in each live show. Harmattan Theater's opus is a cumulative decolonial project in history making. Each performance builds on the ontological directions and cultural meanings generated by the previous performance. The history of Dutch colonization of Lenape lands of what is now New York, for instance, was the original inspiration for the first Harmattan performance, *Henry Hudson's Forgotten Maps*, 2009, Governors Island. Exploring the story of the theft and usurpation of Lenape lands in 1624, this first performance of Harmattan laid the groundwork for a number of ecoaesthetic concerns that are decolonial at their core.

The first decolonial ecoaesthetic practices Harmattan experimented with in *Henry Hudson's Forgotten Maps* were Lenape conceptions of water and land, ownership and hospitality, gifting and potlatch. A performance of over two hours, staged on the pre-1624 geology of Governors Island, colonial cartographies and pre-colonial imaginaries were the primary driving aesthetic of the performance. The second decolonial ecoaesthetic practice explored through this groundbreaking performance was the notion of decolonial duration and of a non-Western visuality. The idea was that movements are stretched beyond the spatial constructs of the Davincian "God's Eye" perspective. Drawing on the principles of the multiple heavens of Angkor Wat in Siam Reap in Cambodia, Joseph applies a Buddhist framework that one cannot catch the entire performance in a single viewpoint. The performance is uncontainable, uncapturable in a singular visual experience. One can only see Buddha's feet as it were, as opposed to his entire reclining body—as in the case of the sleeping Buddha of Southeast Asia, Avalokiteshwra Padhmapani, in Bangkok, Thailand. Drawing on Buddhist traditions of visuality, Joseph incorporated into *Henry Hudson* a third decolonial ecoaesthetic practice where multiple dancers converging from different directions over a period of 75 minutes to create an ensemble performance of slow kinesthetics emphasized temporality as a character in the dance-theater piece. Duration and intuition became processes of unraveling the erased histories of the forgotten Lenape shorelines.

Our preoccupation with embodiment within formerly colonized spaces is a decolonial engagement with repressed feelings and historic silences embedded on the materiality of coastal landscapes. Forts, fishing ports, burial sites, landing docks, slave auction blocks, slave transportation pathways, and colonial trading posts are only some of the sites upon which Harmattan's theater projects have been located. The journey to create an oceanic practice is linked to the consciousness of cultivating a decolonial methodology of gestures, of touch and movement, across spaces loaded with colonial histories of enslavement, colonization, and torture. Drawing local populations to sites filled with psychic and historical trauma is an exercise in facing the historic past through performance. Embodiment becomes a tool for historic transformation, for unhinging ossified narratives.

Method to the Storm

The chapters that follow are exercises in impossibility. First, the impossibility of translating between the living world of weather and elements and the linguistic worlds of signs and meaning. Second, the impossibility of translating into written language and reenacting on the page, through descriptions, maps, and photographs, the experience of site-specific performance. These are productive impossibilities that have sustained our writing and continue to shape and push our thinking about what happens in real time when Harmattan performs. Deploying Shiho Satsuka's (2015) definition of translation as the drawing of one world-making project into another, we can say that our distinct world-making projects across collaborative climate ethnography's performative ecological attunement intersect as stormy systems with the stern graphic containment of prose writing. Each system brings its own elemental activity, its own alignment, its own codes and habitual behaviors, unavoidably shaped and shifted by one another and morphed to (re)orient its equilibria, its directions, its force.

Since Harmattan was founded in 2009, each show, each rehearsal, each conversation has provided both our performers and found audiences with a mediated, highly stylized encounter with the living environment in the present as well as with the ghosts of its past. Our performances are archeological excavations, ethnographies in motion, geological documents. This clash between the right here, right now of an ephemeral present, always already gone, and the lingering weight of the past, may be the third impossibility animating our climate interventions. *Aquatopia* is about climate performance. It is an exegesis on sharing space with human and nonhuman communities. This introductory chapter has offered an overview of Harmattan's "storm as method." In Chapter 2, we draw on histories of Venetian navigation and lagoon culture to propose a comparative lagoon aesthetics linking two archipelagic regions, the Venetian Lagoon and the extended archipelagic region of the Laccadive Sea of India. In Chapter 3, Joseph proposes an immersive oceanic praxis to counter the neoliberal imperatives of late capitalist overdeveloped societies. Considering walking as a participatory (auto)ethnographic method, Varino engages with its multiple lineages across activism, community, and contemporary art in Chapter 4. Joseph examines the effervescent performativity of citizenship in the Anthropocene, alongside its attending ethical and aesthetic challenges in Chapter 5. Finally, Varino elaborates on the possibilities of a radiant ecology centered on wonder and estrangement, informed by Catherine Chin's concept of "historical radiance" (2017), in Chapter 6.

The interludes dispersed throughout offer a respite that also functions as an interruption of sorts. We deploy the genre and format of the interlude to describe and document the liminal space-time of performance via words and accompanying images, focusing on some of the key performances Harmattan has produced since 2009: *Aquatopia* on Governors Island in New York (2017); *Acqua Alta* on the Island of Torcello in Venice (2014); *Mar Português* at the Cais das Colunas in Lisbon (2012); *Far Rockaway* in the Far Rockaways in New York (2013); and *Sea Dike* by the Singel Canal in Amsterdam (2014). Our performances are exercises in impossibility, forcing us to make imaginative leaps in embodying our non/human selves and interacting with our surroundings in less anthropocentric ways, even as we creatively anthropomorphize the lived environments we interact with. Ours has been an aesthetics of collaborative survival, not least as a self-funded performance ensemble working with very limited financial means. Unified by a predictive logic of anticipation and prevention, there is a pervasive sense of impending doom as we endure to such disparate degrees the effects of a global pandemic, ongoing military conflict, the very real threat of nuclear war, all permeated by the relentless forces of flood and drought and tsunami and hurricane and wildfire. Remaining attentive and responsive requires the privilege of access to material resources as despair, indifference, and violence abound. And yet: there might be method to the storm.

Notes

1 For a discussion of the terms "multispecies" and "Anthropocene," see for example John Hartigan "Multispecies vs Anthropocene," *Somatosphere*, December 12 (2014) http://somatosphere.net/2014/multispecies-vs-anthropocene.html/ (last accessed July 11, 2022).

2 Crutzen, P. J. and E.F. Stoermer. "The 'Anthropocene,'" *Global Change Newsletter* 41 (May 2000): 17–18.

Interlude

Aquatopia, Governors Island, New York, USA (2017)

The earliest maps of Mannahatta record the tiny island south of it as a pleasant place filled with nut trees. The archeology of Governors Island reveals a rich history of native settlements over a thousand years. It is a history that is impressive for its wealth of archaeological traces, considering how tiny the original landmass of Governors Island actually was before land fill reclamation extended it towards its current size.

The island is a time capsule of the Anthropocene of New York City over 400 years, in ecological terms. It has gone from dikes, piers, and refilled land to expanded land reclamation and landscape redesigning for a future of rising seas. From flat military naval base to a series of undulating landscapes including hills, farms, berms, seawalls, and expansive parkland, Governors Island has transformed into the largest sea wall for New York City.

Watching the island morph from a flat landscape to one of invented hills and contrived elevations, the importance of this climate adaptation landscape redesign began to impact the shoreline of Manhattan visually. I began creating a public art performance on the new seawall of Governors Island called The Scramble. It is a rambling artificial hill filled with heavy stones arranged in a manner reminiscent of the theater of Epidaurus in Greece—the seats of the theater look upon the city of New York and across New York Harbor. The propelling idea for the project called *Aquatopia* was the destructive hurricane called Sandy that destroyed much of lower Manhattan including Governors Island in 2012. The redesign of Governors Island following Hurricane Sandy addresses many of the urban planning issues impacting New York's island ecologies in the face of rising oceans.

Multiple hurricanes in the Caribbean, the Gulf Coast, and along the coast of Texas, which destroyed Cuba, Barbuda, Dominica, Puerto Rico, Houston, and vast swaths of islands, brought back the terrifying challenges of living by the water. Yet the current government has rescinded infrastructural standards of design for climate adaptation and storm surge across the country. Much of New York City is low lying with the projected hundred-year flood plan well beyond the designated low-lying areas of New York City. Much infrastructural preparation needs to be undertaken and dramatically altered to cope with the realities of New York's geological structure if the next storm is to be accommodated. Governors Island is one physical structure that has emerged as a response to this ecological need. Its presence at the mouth of the East River serves as a barrier to Brooklyn's downtown and Manhattan. The new pleasure island of parks, hills, and pathways is a landscape planned around surge and submersion. To create a performance along such a tenuous site is to visit the precarity of coastal living. The reality of letting the sea in.

Flooding is a way of life for many coastal regions around the world. Planning for flooding is expensive and not politically lucrative for cynical politicians. Consequently, the approach to rising water is salvage not prevention. In the United States this approach has been propelled by outdated laws providing funding for disaster relief rather than prevention.

Aquatopia is being created in the aftermath of multiple catastrophic hurricanes in the Caribbean and the Gulf of Mexico. The widespread destruction of entire islands is unfathomable. *Aquatopia* is a process of addressing the questions of living by the sea. How can we better prepare our selves for such a large scale calamity? What needs to be done to prepare for the eventuality of the next major storm? *Aquatopia* probes the fragile coexistence of humans, nonhuman forms, and nature. Set against the dramatic landscape of an artificial hill, the performance asks the audience to consider the meaning of living on islands by the sea. Facing the island of Manhattan, the performance draws the audience into an immersive engagement with the landscape itself. Walking, climbing, sitting amidst the scales of stones, the audience is actively engaged with the process of performance. Actors move amidst the audience, drawing them into the poetic space of meditating on the nature of the sea. We are presented with Dante Alighieri's dark vision of the tempestuous sea and its violence wrought upon the fate of men. Reciting from the Fifth Canto, James Cascaito offers a glimpse inside Dante's liquid world, filled with drowning unrequited sinners.

Heinrich Heine's immortal Lorelei, performed by Lisabeth During, delights the audience into a hypnotic trance of deadly consequences. The siren beckons lost souls towards the deep waters, only to drag them into the undertow of the troubled ocean, discernible at the narrow lip of the famed Verrazzano Narrows, opening up the vista to the Atlantic. Sofia Varino's Fernando Pessoa evokes the Portuguese Sea that shaped the navigational empires of the sixteenth century in all their violent, deadly splendor. A cavalcade of ruins rise with the poem, a landscape of forts and ports, of which Governors Island is a point among them. It is noted that some of the first Africans to arrive in New Amsterdam were Portuguese-speaking Dutch slaves from Pernambuco, Brazil, brought all the way from Angola (Berlin and Harris, 2005). Pessoa's *Mar Português* reminds the audience of the historic origins of Manhattan's Atlantic slave history as being part of the vast "Portuguese Sea." Its legacy of slaves and spices is enmeshed among the cobble stones of seventeenth-century Wall Street. Following Pessoa, the poet Frederico Lorca (performed by James Cascaito) mournfully sings of the violence of water.

Amidst the voices of the poets across the seas, the metaphoric Hudson River swells in its undulation, wrapping the new seawall in its angry flows, fluttering in the wind its uncontainable sweep of fabric. The performance used Indian saris in colors of the Hudson River, to create the scales of the river swaddling the new mountain that has been created as New York's new berm or seawall. The island's solitary presence at the mouth of the harbor leaves it standing alone, a sentinel too little, too late in the ecological reimagining of New York City's island ecology.

Aquatopia was an immersion of New York's community in the frailties of changing coastal climates. Raging fires in California, deluged coastlines along the eastern seaboard, and the heating of the planet are felt all across the continent. A foreboding has settled within the coast, yet the government's actions are irresponsibly opposed to the reality of the plight on the ground. In Puerto Rico, water-deprived victims of the devastating hurricane are drinking water from toxic waste sites. As shorelines erode and the tenuous reality of living by the sea becomes more precarious, an intensity of the everyday will transform how humans engage with the coastline. Performance will increasingly become a necessary means of drawing people to the shoreline, enabling creative engagements with its possibilities and its perils, its liminality and its limits.

For documentation of *Aquatopia* (2017), please visit harmattantheater.com.

2 Multidirectional Thalassology

Comparative Lagoon Ecologies

May Joseph and Sofia Varino

What are the historic relationalities between the Venice lagoon and the Indian Ocean? How has the *acqua alta* flooding of Venice, accompanied by the mnemonic histories of the Venetian lagoon, impacted understandings of lagoon cultures in the global South, particularly the Malabar Coast of South Asia? Drawing on histories of Venetian navigation and lagoon culture, in this chapter we propose a comparative lagoon aesthetics, one that would link two archipelagic regions, the Venetian lagoon and the extended archipelagic region of the Laccadive Sea of India. While we believe a contemporary archipelagic study connecting these two regions does not currently exist, the historical archives suggest otherwise. We draw on the Venetian Camaldolese monk and cartographer Fra Mauro's *Mappa Mundi* from the fifteenth century to initiate this comparative dialogue between North/South island ecologies, seafaring histories, and ocean futures affected by climate change and rising sea levels.

Venice, Postponed

In May 2020, Harmattan Theater was scheduled to do a performance on the Devil's Bridge (Ponte del Diavolo), Venice's oldest stone bridge, without a handrail. It was to be the repetition of a performance that we, May Joseph and Sofia Varino, had created together on the Bridge at sunrise on a sunny November morning in 2014. This repetition was to accompany a talk for the "Living, Narrating and Representing Venice and its Lagoon" conference. Sometime in early February, as concerns about COVID-19 and the coronavirus pandemic increased in Italy and across much of Western Europe, the conference organizers sent out email messages canceling (well, postponing) the event. Within a few days after that, COVID-19 had gone viral in every possible sense, and what came to be called "cancel culture" had become pervasive,

DOI: 10.4324/9781003359906-2

with social life moved online as Zoom and other similar platforms gradually became the main means of communication for so many.

Our planned talk and performance in Venice were meant to contribute to a historically informed empirical phenomenology of climate change as it impacts water-bound ecosystems, considering the circuitry of urban infrastructures in a city like Venice. As such, what we often referred to as *Acqua Alta II* or *The Venice Project* was meant to contribute to the fields of environmental philosophy, decolonial history, climate aesthetics, as well as embodied phenomenology and artistic research. Soon after the conference was canceled, the COVID-19 pandemic came to occupy a more and more central role in everyday life, particularly for those living in high density urban areas, where lockdown measures eventually came into effect. Once the links between the pandemic and climate change, deforestation and habitat loss, became clearer, as shown in a steady stream of news and scientific articles, the viral contours of climate change structured our daily embodied existence wherever we were, affecting marginalized communities and especially impacting racialized, disabled, and ageing minorities. The novel coronavirus had found a way into our collective non/human corpus, coming to literally inhabit far too many individual bodies with disastrous consequences. At the same time, the global Black Lives Matter movement gained momentum, further establishing networks beyond its solid foundation as a USA-based movement, instigating and supporting racial justice activism on a transnational scale. As our "Venice Project" receded further into the background of managing the daily chaos of that second quarter of 2020, new questions came to occupy our minds and shape our thinking around climate change, maritime histories, archipelagic geographies, and the new environmental configurations of our daily lives. How can we respond, how can we participate, how can we be present in and through the multiplying crises of our times when we are required to stay safe, remain distanced, avoid physical contact, and seek protection within the walls of our homes or behind masked faces? We remain exposed within the bounds of our confinement, and becoming acutely aware of our own and everyone's vulnerability is precisely what can help us navigate our troubled pandemic times. A multidirectional thalassology began to emerge as our method for thinking capaciously across all these lines of flight.

Venetians in the Malabar

At the center of a dimly lit room in the Biblioteca Nazionale Marciana hangs the magnificent *geographicus incomparabilis* (Braudel, 1981: 407) of Venice, Fra Mauro's 1450 *Mappa Mundi*. Scalloped at the top of the blue and gold canvas with turquoise and white waves lies the archipelagic global South. Redolent in hues of green, brown, and red, the far-flung islands of the Maldives, the Andaman Islands, the Malabar Coast, and the archipelagoes of Malaysia and Indonesia occupy the upper third of the map. This region, according to historian Hiram Woodward (2004; see also Acri, 2018), an "integral unit" to Buddhist flows into the ninth century, included the Nusantara Archipelago (Manguin, 1993, 2011).

Delicately adorned floating islands from South Asia to the South China Seas structure *Mappa Mundi*'s startling upside-down orientation. The Malabar coast is prominently marked on it (Joseph, 2019a). Joseph gasps when she finally beholds this extraordinary map in Venice, the reason for her visit. A swirl of seas, the *Mappa Mundi* positions today's East Africa, India, and Sri Lanka in its geographical north. Fra Mauro's rendering of the world was influenced by Ptolemy's *Geographica* as well as the travels of Marco Polo and Nicola Conti through the Malabar region of Koulam (Quilon), Cochin, and Calicut (Bracciolini, 2004; Polo, 2003). The *Mappa Mundi* is one of the main historical sources for devising the site-specific performance *Acqua Alta* (2014), directed and choreographed by May Joseph, and performed by Harmattan performers Sofia Varino and Marit Bugge, on Venice's earliest inhabited mudflat, Torcello.

Fra Mauro's decentering *Mappa Mundi*, gilt with lapis lazuli blue, vermilion red, and gold, is a testament to what Fernand Braudel describes as "what the world looked like to an Italian in 1450" (Braudel, 2019: 29). In the period that Braudel (2014) delineates as "the long Middle Ages," spanning the Medieval to the Renaissance era, Venice was an archipelago whose origins lay in the history of flight and refuge from its mainland. Its glory as a trading city hinged on its knowledge of the land and sea routes to Asia during the medieval period (Braudel, 1995). The sources of the Murano-born Mauro's *Mappa Mundi* were rooted in the global travels of the two Venetian travelers, Marco Polo and Nicola Conti, from China to India through the Middle East and back to Europe, among other archival references such as Pliny, Ptolemy, and Abyssinian visitors to Florence (Cataneo, 2011).

The fifteenth-century visually disorienting *Mappa Mundi* allows island theorists to ask questions about comparative geographies in startling ways. It stages the tension between what Bruno Latour calls the "terrestrial" and the "planetary" (Latour, 2018). Latour argues that the scales of our conversations on climate have been deflected to the Sirius point of view. We speak of the planetary while looking downward from the heavens, losing sight of terra firma from the point of view of the earth beneath our feet and the world around us. Fra Mauro's planetary orbit captures this escalating dialectic between the terrestrial and the planetary from the vantage point of a decentered, small island perspective, that of fifteenth-century Venice. It privileges the marine sphere as the planet's foundational materiality. The inverted *imago mundi* documents information on nautical, navigational, and commercial knowledge for Western epistemic frameworks in its cartouches and legends. The *Mappa Mundi* depicts the world as "a sea of islands" (Hau'ofa, 1994: 1), drawing the spectator into a geopoetics of islands (Balasopoulos, 2008) and situating Venice at the center of a comparative new thalassology, an emerging archive of the seas preceding Vasco da Gama's sea route to India.

In their influential article "The Mediterranean and 'The New Thalassology,'" Peregrine Horden and Nicolas Purcell state: "The systematic comparison of real and metaphorical seas can suggest a new configuration of history, and one that might attain a global scale. So promising, indeed, does the notion of a sea or an ocean appear for this task that the term 'the new thalassology' has seemed an appropriate coinage to denote it" (Horden and Purcell, 2006: 723). Fra Mauro's *Mappa Mundi* occupies a historically significant position in the emergence of such a critical study of sea and ocean epistemologies foregrounding islands. It presents a Venetian world view of an island-city that was part empirical and part mythic. Venice was a pivotal center in the convoluted history of sea empires. Its maritime identity, however, was always situated alongside the unrecorded histories of multiply dispersed ocean journeys across the global South. This hidden history signaled by the few diaries left by navigators such as Marco Polo, Nicola Conti, and Ludovico de Varthema gesture to the multi-sited repercussions of La Serenissima's global reach. Such a dispersed history of Venice's islandness necessitates imaginative detours across its multilayered navigational pathways. According to Angelo Cattaneo, Fra Mauro's *Mappa Mundi* is arguably the most comprehensive surviving document of how the world was imagined from the perspective of a Venetian in the fifteenth century (Cattaneo, 2011).

To embark on a study of comparative island geographies requires innovative research methods. One approach towards a critical comparative study of a maritime city such as Venice is Michael Rothberg's concept of multidirectional methodology (Rothberg, 2009). Rothberg offers an invigorating way into comparative geospatial thinking. He argues that to fully engage with the junctures of minor, erased, or vanished histories, the archivist has to forge multidirectional linkages. The violence of colonial pasts necessitates unexpected detours in the construction of a working archive. One has to interrogate *what counts as archive* using a transversal approach that cannot be located in any single site, region, or genre of documentation. Rothberg invites the comparative historian to combine empirical history with other forms of accounting for the past. Memory for Rothberg is key to opening up the spheres of multidirectional research (Rothberg, 2009).

Rothberg's methodological suggestions liven up the literary and cartographic possibilities posed by Fra Mauro's *Mappa Mundi*. The world in his map is a relational one of islands linked through the accumulating memories of cartographers and travelers. Attesting to the deep historic, ontological, and aesthetic attunement connecting Venice to the oceans of the world, the *Mappa Mundi* captures projected uncertainties about the nature of *mare incognitum*, the unknown sea, with Mare Indicus embodying the apotheosis of the Mediterranean seafaring imagination. As Jacques Le Goff observes, the Indian Ocean in Fra Mauro's time was impossible to reach and in the thirteenth century was a place beyond Europe's imagination. It was a sea that lay on the outer edges of the known and the unknown worlds, a liminal place of fables and terror (Le Goff, 1980).

A recent reading by Angelo Cataneo shows how Mauro's *Mappa Mundi* offers multiple perspectives on the Ocean: "The *mappa mundi* describes vast marine spaces. The seas represented in the *mappa mundi* are part of the *oikumene*, in the sense that they are navigable and their thousands of islands are either inhabited or inhabitable" (Cataneo, 2011: 118). In particular, Mauro suggests an "'openness' of the Sea of India through the circumnavigation of Africa, so that it communicates with the western sea, the Sea of Darkness" (Cataneo, 2011: 117). Cataneo draws on the following inscription by Fra Mauro on a *Mappa Mundi* cartouche to bolster his point:

> Some authors write that the Sea of India is enclosed like a pond and does not communicate with the ocean. However, Solinus claims that it is itself part of the ocean and that it is navigable in the southern and south-western parts.
>
> (Cataneo, 2011: 118)

According to Cataneo, Fra Mauro's inscriptions on the *Mappa Mundi* are suggestive of the projected fantasies and fears that Mare Indicus invoked for the Mediterranean before the opening of the sea routes to India around the Cape of Good Hope. Mauro was clearly influenced by Nicola Conti's descriptions of the Malabar Coast as a place of serpents and strange amphibious creatures. Conti's diaries describe flying cats and winged serpents in Koulam (Quilon) and he writes of monsters in human form that are fish-like and come forth at night in the water (Major, 2010). The Mare Indicus in Mauro's *Mappa Mundi* was a place of untold riches, nocturnal terrors, and turbulent seas (Cataneo, 2011).

A Malabari in Venice

The *Acqua Alta* project was conceived as a theoretical engagement with decentering the Mediterranean world on historically new terms, through an embodied phenomenology of the sea. It is a performative experiment in micro-history. As a performance-based artistic research project, it deploys queer embodiment techniques to explore that which lies outside the conventional narratives of Portuguese and Venetian maritime histories. Embodying the hidden memories of La Serenissima through lived presence, *Acqua Alta* asks forgotten questions about the archipelagic histories of minor seas such as the Laccadive Sea, located outside the imaginary construct of Europe's center, Venice, yet so prominent in Fra Mauro's rendering in the fifteenth century. Through Mauro's *Mappa Mundi*, the performance excavates forms of knowing that fall outside hegemonic readings of Venice's history. *Aqua Alta* achieves this through a series of mobile, speculative encounters around the Venetian lagoon as a means of marking the *carreira da India*, or sea route to India (Cataneo, 2011: 124).

Acqua Alta is a performative extension of Fra Mauro's precise and fantastical map. It is a staging of the histories of human habitation between the Indian Mediterranean of K. N. Chaudhuri (Chaudhuri, 1985) and the Classic Mediterranean of Fernand Braudel (Braudel, 1995). Creating the performance *Acqua Alta* on the Venice lagoon is a way of marking the multidirectional traces, memories, and materialities that shape the most iconic of lagoons in the most historicized sea in the world, the Mediterranean (Horden and Purcell, 2006). For Joseph, *Acqua Alta* is an open work (Eco, 1962) examining David Abulafia's notion of the Mediterranean as "the history of human encounters" (Abulafia, 2005: 68). As Abulafia writes, "[it] is not just the history of what happened on the sea, but the history of the way the inhabitants of the opposing shores of the sea interacted across the sea" (67) that needs to be rethought. Traveling from the farthest shores of Mare Indicus, from the lagoon of Quilon in the Malabar Coast, to create this performance on the Venetian lagoon is for Joseph an experiment in the encounters between multiple histories of the seas. It is a processual probing into the navigational histories of Venetian travelers moving among the Laccadive Sea, the Arabian Sea, the Indian Ocean, the Cape of Good Hope, the Atlantic Ocean, and the Mediterranean Sea into the Adriatic Sea and the Venice lagoon. Traveling from the Malabar coast to perform in Venice is an exercise in multidirectional thalassology. It is an embodied excavation of multiple seas, a dialogue between historically connected marine spaces.

Horden and Purcell write that "sea and ocean history is more novel than it sounds ... its scope and its methods are so distinctive as to make it an exciting – and quite unpredictable – area of reflection and research" (2006: 722). For Joseph, this journey from the hemispheric South to the hemispheric North stages Fra Mauro's cartographic inversion. It performatively foregrounds "what matters – what makes these seas new and exciting historiographical categories – is the density and variety of human connections across them" (Horden and Purcell, 2006: 739). *Acqua Alta* participated in this historical rethinking and contributed to an expanding archive of ocean histories through a return of Venice's own repressions in the form of two historically bound social subjects: Joseph, a Malabari living in New York, and Varino, a queer Portuguese Jew formerly based in New York and now living in Berlin. *Acqua Alta* conjoins histories of colonization with decolonial reworking of embodied memories through the iconic crossroads of Lisbon, Venice, and Quilon, which Conti describes as one of the biggest ports in India in the fifteenth century.

"Venetian Gulf"

The 2014 *Acqua Alta* performance took place in the Venice lagoon and involved an hour's boat ride from Cannaregio to Torcello, stopping at Murano and Burano en route to the lagoon. For Joseph and Varino, the boat journey was part of the process of the lagoon performance. Its duration brought to the fore the multiple imperial thalassographies of Mare Indicus that haunt the Venetian lagoon. The duration of the vaporetto ride invoked the seafaring empire of Venice as multidirectional methodological praxis. The Venice lagoon is a liminal place encapsulating the merging of the Indian Ocean with the Mediterranean Sea. The lagoon functions as a node where the shared aquatic ecologies, histories, and cultures of the two island regions come together, intermingling. The lagoon, known to medieval writers as the "Venetian Gulf" (Abulafia, 2013) invokes the primal space of Mare Indicus depicted by Fra Mauro, while also suggesting a more interconnected lagoon aesthetics (Le Goff, 1980; Finch, 1983). Venice's lagoon was one of the crucibles of the world's crossroads in the fifteenth century. To look into the lagoon's waters is to imbibe the rich, often violent and deadly, intersectional histories of the Silk Road, the Spice Routes, the Indian Ocean trade routes, and the era of colonial expansion into the Southern hemisphere. Global trade through land and sea routes transacted through the city of Venice en route to Istanbul. Likewise, the impact of Venice on the outer reaches of the then geographies of the Southern hemisphere were multiple but less documented, as the Portuguese and the Dutch dominated the many colonized entrepôts of Asia and Africa by the 1500s (Le Goff, 1980). However, journals and reports of Venetians in the Malabar Coast remain among some of the more detailed coastal and oceanic descriptors of seafaring knowledge of the Malabar from the period (Major, 2010; Polo, 2003; Varthema, 1997).

Drawing on the layered histories of the Indian Ocean, Portuguese maritime records, and Dutch sources, Joseph and Varino reach across the oceans from the fifteenth century to retrieve Venice's multiple mnemonic markers of the global South that have since been erased or forgotten. The origins of Fra Mauro's map is one such reminder, where the Maldives, the Andaman Islands, and the Malabar Coast of India hold the geopolitical, affective, and material power of being located towards the visual north of the canvas. The waters between the Maldives, Ceylon, and the Malabar in Fra Mauro's map refer to the Laccadive Sea, lagoon-like in its cartographic depiction between the Indian Ocean and the Bay of Bengal.

Reading Mauro's map from the frog's-eye perspective of the Malabar Coast while traversing the Venice lagoon, Joseph speculates about Le Goff's notion that Fra Mauro renders Mare Indicus as a closed space. For Joseph, Le Goff's observation presents a metaphysical tension in the *Mappa Mundi* between the open sea outside Venice's lagoon and the vastness of the closed ocean represented by Mauro at the top of *Mappa Mundi*. According to Joseph, Fra Mauro's map is not so much closed as Le Goff suggests but rather depicted as lagoon-like. Fra Mauro's rendering of Mare Indicus projects a lagoon aesthetic into the ocean that captivated the imagination of Medieval Europe. Joseph suggests that Mauro's map projects a material and metaphoric connection between the lagoon culture of Venice and its Malabar coast influences from the thirteenth to the sixteenth century. The Laccadive Sea becomes a scaled-up Venetian lagoon, with Taprobane (Ceylon), the Malabar, and the Maldives framing the centerpiece of the Andaman Islands. This lagoon aesthetic provides the philosophical thread for our performance on Torcello Island in 2014.

Multidirectional Thalassology

In the thirteenth century, Marco Polo visited Quilon along the Malabar Coast in India and in the fifteenth century Niccola Conti stayed in Quilon, as did Ludovico de Varthema subsequently. All three explorers wrote of Quilon in ways that invoke Venice. The site of Conti's and Polo's visit, Quilon, is the archipelagic lagoon town where Joseph and her mother grew up on the Ashtamudi Lake, similar in structure to the Venice lagoon. Polo (2003) referred to Quilon as "the residence of many Christians and Jews, who retain their proper language" (377). It is a lagoon culture of interconnected small islands open to the Arabian Sea through barrier islands. The city of Quilon is located inside the lagoon of Ashtamudi, amidst other barrier islands similar to the Lido of Venice. Consequently, Quilon made a perfect military port, as Polo observes in his diaries. It was the primary base for the Portuguese and later the Dutch in the Southern Hemisphere, as it was located on an archipelago with immediate access to the Indian Ocean. There was something Venetian about the geology of Quilon, an archipelago protected by barrier islands and the first safe harbor coming across the Bay of Bengal from Malacca. Coming from Portugal, it was the last big port leaving the Malabar across Cape Comorin towards Aceh. This was why in the fifteenth century the Portuguese and later the Dutch set up Quilon as the largest colonial port in South Asia at the time. It was a place where all travelers had to stop by to refuel en route to Africa from Asia. It was also strategically protected by islands and treacherous currents (Varthema, 1997; Major, 2010; Polo, 2003).

Marco Polo (2003) wrote expansively about the "multitude of islands in the Indian Sea" (397). He observed: "I have heard, indeed, from mariners and eminent pilots of these countries, and have seen in the writings of those who have navigated the Indian seas, that they amount to no fewer than twelve thousand seven hundred, including the uninhabited with the inhabited islands" (397–398). This lagoon region of island ecologies on the Laccadive Sea that Polo invokes includes Quilon, Zeilain (Ceylon), the Maldives, and the Laccadives. This medieval seafaring connection between the lagoon of La Sereníssima and the Malabar archipelago, with its Italian traces from Ostia (Roman trade) to Venice (Polo, Conti, and Varthema) opens up the influence of Venetian navigation on Indian Ocean networks and vice versa (the impact of Malabar Coast spices, silks, and coastal ecologies on Venetian culture) which were considerable during the medieval period (Abu-Lughod, 1989; Braudel, 1981, 2019; Chaudhuri, 1985; Das Gupta, 2004; Subrahmanyam, 2004; Vadakkekara, 2007).

Revisiting the medieval histories of thalassology unpacks the hidden networks of modern ocean-bound thinking. The turbulent and dilated travels of Venetians like Marco Polo and Niccola Conti informed cartographic mappings of the global South. The archipelagoes of the Laccadive Sea region are given prominence in Fra Mauro's map in a manner that surprises the contemporary viewer. What catches the eye is how the archipelagoes of the global South were deemed of cartographic significance to the maker of the map, from his island sanctuary of S. Michelle di Murano.

In the era of climate change and the much contested "Anthropocene," Venice's lagoon culture and its history of *acqua alta* mirror the Laccadive Sea as a lagoon around which an extended archipelagic imaginary contends with rising oceans. Varino and Joseph's 2014 performance on the Devil's Bridge in Torcello explores these connections. When we return to Venice to continue our planned performative ethnography, our project will be threefold: to deepen our study of the Venetian lagoon in light of the Venice storm surge of November 2019; to probe the ecologies of Venetian navigation; and to create a performance piece that engages with our preoccupations across global lagoon cultures: Jamaica Bay in New York, the Laccadive Sea, and the Venice Lagoon. A desire to learn from and interact with the Venice lagoon is the core catalyst for our project. The Venetians have been dealing with the challenge of storm surges for quite a long time now and continue to struggle with inescapable damage. San Marco was flooded as we prepared for our performance in 2014, with the gangways raised above the water in the piazza, demonstrating the provisionality of Venice's own reckoning with its unfolding calamity. Our performance, in solidarity with the quintessential island city of the world, is a marking of Venice's vulnerability and painful resilience. What happens in Venice will eventually happen in other low-lying regions like the Malabar Coast. How can Venice help us prepare for the inevitable coastal precarity impacting much of the global South? What are the lessons the world can learn from Venice? These are the questions that drove the initial 2014 *Acqua Alta* project.

Lagoon Ecologies

The COVID-19 pandemic encapsulates many of the current crises and consequences of our times: ecological devastation, scientific uncertainty, bio(in)security, and brutal disparities in access to healthcare. We are now living in an age of social distancing, perhaps more accurately described as *physical* distancing. Island cities like Venice with its lively lagoon ecologies have experienced these restrictions in distinctive ways. What has it meant for Venice's island city to be locked inside its own lagoon during the COVID pandemic? Venice's lagoon dwellers have experienced particular hardships, redefining how they socialize in open, public spaces in a culture dependent on and driven by the tourist economy. The future of Venice's lagoon landscape will demand creative and improvised use of waterfronts, shorelines, plazas, parks, and streets that are closely aligned with Harmattan aesthetics. For island cities like Venice, the isolation of many months of quarantining and social distancing has been extremely challenging both economically and culturally. As commercial flights were canceled and boat services and water-bound transportation scaled back, the new implications of what it means to be La Serenissima in the age of COVID are beginning to emerge. The tools and strategies Harmattan Theater deploys in its environmental research are informed by an ethos of uncertainty aligned with experimental forms of knowledge production that island cities like Venice and New York will have to draw upon, especially after COVID.

As an environmental performance ensemble, Harmattan is oriented towards the lived environment as the necessary condition for art, and indeed all living things, to emerge. The case of Venice, with its body of water combined with the phenomenon of *acqua alta*, or seasonal flooding due to higher than usual water levels, is unique across the Mediterranean. Venice's high water is an urgent problem. No longer merely confined to the winter months, flooding regularly inundates the city, creating a devastating ecological bind for the vulnerable archipelago. Some of Venice's most pressing challenges include how to manage the flow of the Adriatic into the Venetian lagoon, how to cope with rising ocean levels, and how to adapt a transformed human-made landscape of mudflats into a vision for the future.

Our phenomenologically and historically attuned exploration of Venetian aquapelagic geographies links the phenomenon of *acqua alta* to other sites of flooding and water-bound precarity like New York City, Lisbon, Amsterdam, the Maldives, and the Malabar Coast. The deep connections of the Venice lagoon to the Laccadive Sea region are historical, geographical, and ecological. One of our points of departure is that contemporary studies of lagoon and island cultures tend to be so localized that they omit the vast global networks across which aquatic aesthetics and geographies elsewhere in the world were historically shaped by European navigation. The relevance of the Venetian lagoon for an ecology of precarity across the global South is of considerable importance. What happens in Venice will happen, and in many instances is already happening, across the global South, although it may currently draw less attention. This is why it is urgent that communities across island cities of the world collectively participate in the interconnected present and future of their water-bound sites as performative ecologies.

At the core of Harmattan's approach to examining geographical space ecologically has been an insistence on the multiple temporalities of performance in real time colliding with historical temporalities. In this liminal space between historical past and present, a tension begins to emerge (Joseph, 2020). Our embodied engagement with the Venice lagoon and our approach to researching islands, archipelagoes, and coastal areas is ethnographic, participatory, immersive, and performative. We engage with Venice through site-specific environmental performances, immersing ourselves in the outer islands to study the waterlines and find out what they can tell us about Venice's lagoon life today, linking Venetian navigational routes to Dutch and Portuguese ones. Harmattan's *Acqua Alta* project catalyzes the (un)learning humans must undertake to reckon with the volatile nonhuman forces and colonial histories haunting archipelagic choreographies.

Aquatic Activity and Coastal Performatives

According to Jonathan Pugh (2018), the relational and archipelagic turns in island studies have forged a sense of interconnection and "foregrounded how we live in a world of relationality rather than 'static' islands of the world" (94). Drawing on this notion of interconnected relationality enables us to articulate how the agency of Venetian waters extends well beyond their local lagoon geographies, performatively rippling over to aquatic activity alongside the Malabar Coast and the global South in unexpected entangled (re)enactments. The interconnection at the core of Venice's fluid geographies gestures towards a viral mode of exponential transmission whereby streets, islands, streams, bodies, objects are continuously touching, where contact is non-negotiable, air is shared, and airborne droplets may carry enough viral load of genetic material to generate an immune response.

The aquatic life of Venice is lived underground as well as on its visible, gutted surface. The seasonal floods tie into a much vaster network of climacteric activity, whereby the city becomes a semi-aquatic environment. Oceanic histories dispersed across colonial networks have fostered incommensurate, rapidly shifting geographical and temporal scales. At the nodal intersection of postcolonial studies, political ecology, experimental geography, and island studies, Harmattan Theater has devised a transdisciplinary method for situating bodies in real time in relation to these large scale geographical and temporal frameworks, using the medium of site-specific performance to demonstrate the implication of distant times and places in the lived materiality of climate change events.

Writing about Haiti in particular, in *Island Futures* (2020) Mimi Sheller deploys the phrase "coloniality of climate" (8) to propose deep and vast connections between climacteric phenomena and the colonialist projects of modernity, encapsulating the many ways in which (neo)colonial forms of violent and exploitative extraction come to produce the very conditions of social inequality and environmental degradation that then amplify apparently "natural" disasters like earthquakes, floods, storms, or volcanic eruptions. In *Queer Phenomenology* (2006), a phenomenologically attuned analysis of embodied experiences of racist and sexist violence, Sara Ahmed mobilizes the concept of "orientation" to formulate the ways in which geopolitical and historical structures become sedimented over time in our everyday lives, coming to shape how we occupy space and move through the world. In *Acqua Alta*, we similarly sought to reorient ourselves toward older mappings and networks of water-bound environmental practice, excavating the multidirectional topographies of the lagoon to reenact them across the Devil's Bridge in Torcello. Performatively engaging with these networked systems, Harmattan has continuously returned over the course of a decade of environmental art to the rich theoretical arsenal of phenomenology for a materially oriented, situated account of historical processes whose sheer magnitude cannot be fully grasped without aesthetic involvement. The phenomenological focus on the sensory and the kinetic alongside the linguistic has informed the aesthetic experience of a Harmattan performance for both audiences and performers.

From the fluvial conditions of Venice's geographies to the economic exploitation necessary for the enduring success of colonialist capitalism, the conditions are always impossible. Thinking about *acqua alta* and flooding requires flexible modalities of thought and action that encompass resilience and endurance as well as fragility and vulnerability. We are permeable, water-made, and water-bound. Even for those residing far from ocean shores—as we have seen with the snowstorms and power cuts in the US state of Texas in February 2021—climatic conditions, water consumption, and energy resources regulate and permeate the commonality of our living bodies, with fluids circulating among our internal viscera and mucous membranes and orifices interacting with and filtering solid, liquid, and airborne particles. The aqueous, interstitial capacities of water are a precondition for the emergence and sustenance of planetary life as we know it. Around the lagoon of Venice, water acquires an imminent urgency sounded by the siren bells at dawn during *acqua alta*.

Bearing the weight of its historical past and of its status as an iconic "sinking" city and tourist destination, the center of Venice in 2014 was so filled with the presence of human sociality that we had to find an alternative performance site, looking for the reverberation of its dense history at its geographical margins, dispersed and distributed. The formative impact of Torcello in the emergence of Venice guided our choice of the island as our performance site. We made our journey there as a group of performers ready to improvise, collaborating with human and nonhuman life forms, with the elements, the built environment, and the climacteric conditions of the day. As we moved further into the island looking for the famous Ponte del Diavolo, the architectural minimalism of the design proved a source of further estrangement and enchantment. The solid stone structure without railings provided us with a twelfth-century open stage charged with history and filled with ghosts. Thick with stories, it invoked an uncanny liveliness we could not resist as we crossed it, an ensemble of bodies moving in tandem, deploying the meditative slow walk of Harmattan performances and following the improvised musical arrangements of Girardi's blues folk guitar. As Joseph and Varino have previously observed:

> Venetians have dealt with the challenge of storm surge for quite a long time now and have managed to accept its inescapable damage yet remain flexible in the face of ensuing calamity. This performance was a marking both of their vulnerability and remarkable resilience.
>
> (2017: 162)

It also enabled us as an ensemble to performatively interact with the volatile qualities of water, its agentic activity, and the myriad ways in which human and nonhuman populations have historically tried to harness and/or control its power to both sustain and annihilate life in any given ecosystem.

Along the lines of Jane Bennett's political ecology of things in *Vibrant Matter* (2010), Harmattan prefers tales where the agency (understood here to encompass *both* activity and passivity) of matter and the visceral materiality of human lives takes precedence over the supposed mastery of "human" action and reason. The ecosociality of

Harmattan's performance projects suggests in its methodological and aesthetic directions an engagement with a viral logic of contact, transmission, relationality. We are performing contagion in how we come to invade and occupy specific sites, how we collectively relate and transform, how we respond to immediate circumstances by fostering ties and relating to others. This is, of course, a metaphorical manner of considering how the novel coronavirus and the COVID-19 pandemic has forced us to be far less physically enmeshed while demanding an imaginative effort of participation and presence in other, perhaps digitally mediated or distanced, but nonetheless observant, responsive, lively ways. Harmattan's environmental performances are closer to landscape art than theater, closer to large-scale installation than performance art. Our site-specific performances are conceived as political, geographical, philosophical, and aesthetic interventions in public space and the lived environment. We seek to performatively engage with the colonial past not to mechanically reenact it but to tentatively begin to acknowledge its ghosts, its perpetual haunting, its inevitable, brutal hold on the non/human lives that can be lived in the present. In a Harmattan project, performance then functions as an interruption, affording a way to physically and cognitively occupy the embodied present, both in a spatial and in a temporal sense. Performance provides us an aesthetically and socially attuned method for multispecies collaboration in public space. It is a practice of historical (re)enactment with a difference, oriented towards a more democratic, inclusive, accessible future, livable for all rather than for a selected few.

Learning from *Aqua Alta*

In his persuasive book *If Venice Dies* (2015), Salvatore Settis asks unsettling questions: What is a Venice without Venetians? If the soul of a city is its people, then what does it mean to have a city without a soul? Settis is hard-nosed about his questions, observing that the lagoon of Venice is the ecosystem that maintains the city's sustainability. The lagoon was Venice's countryside. It was also the city's extended natural resource, where vegetables, fruit, fish, and salt were derived and which nourished it. Venice's lagoon was an interdependent ecosystem of monasteries, hospices, leper colonies, boathouses, and spaces for alternative industries that supported the peripheral island economies of the city. The systematic decimation of the human and nonhuman cultures of the Venice lagoon through the slow emptying of the city's population over decades has destroyed this delicate ecology, Settis documents. Settis highlights the contradictory tensions between the erasure of Venice's culture as Venetians have known it, and the escalation of Venice's real estate value in the distorted market of international speculative economies. This Venice that has been emptied out of its Venice-ness is a place of loss. It is this mournful habitation, this site of haunting, that Joseph and Varino summon in *Acqua Alta*. As the human crisis exacerbates an ecological catastrophe, the soul of Venice is at stake.

This ongoing crisis at the heart of Europe illustrates what Bruno Latour calls the relationality of the terrestrial, an embodied situatedness at a human scale that is earthbound rather than merely planetary. It is a relational site already proposed by Fra Mauro in the fifteenth century as a marine space where the "thick" waters of the sea differ from the "thin" waters of the rivers (Cataneo, 2011: 115). In Fra Mauro's time, the understanding of the rhythm of the tides and "the delicate balance between land and water and the phenomenon of the 'boiling of the waters,' had special significance" (ibid.). Cataneo notes that the lagunar city of Venice was one place where the volatile tides were most evident during Fra Mauro's time. These "boiling waters" now persist as the catastrophic phenomenon of *acqua alta*. It is a condition of inundation, of submergence, that connects La Serenissima once again to its historical fascination with the islands of the Indian Ocean. Only this time, the archipelagic connection is bound to climate change and rising sea levels. The conversation this time is not about that which is unknowable, but rather that which is terrifyingly predictable, the progressive warming of the oceans.

Why do we insist that it is important to think about the similarities between the marine ecologies of the Laccadive Sea and the Venetian lagoon? One reason is that if we are to understand the shared commonalities and find solutions to grapple with our current environmental crises, efforts need to be made to dispel the illusion of a radical distinction between the global North and the global South, especially regarding the impact of climate on a global scale to be understood in terms of a terrestrial relationality, as Latour proposes (2017). The Laccadive Sea ecology is one that is ringed by thousands of islands strung along the Laccadive Islands including the Maldives, Sri Lanka, and the near shore archipelago of the Malabar Coast. Joseph has conducted fieldwork across this region for the last two decades, studying local coastal communities in Kerala and the Maldives. Her conversations with fishing and shoreline communities, coastal residents, and non-governmental actors underscore the infrastructural differences between hemispheric climate preparedness.

There is a deep bifurcation between the large-scale technological solutions to climate threats that are currently being developed and implemented in the global North, and the fragile accommodations undertaken by coastal and island communities in the global South. The infrastructural approach to managing climate in the global North centers around adaptation (Bush et al., 1996), raising habitats and coastal embankments, (re)imagining flexible amphibious environments, and bolstering sea level infrastructure for the 100-year and 500-year floodplain (Keith, 2017). For regions around the Laccadive Sea, particularly along the Malabar coast, it is primarily one of low-tech adaptation and mitigation. Coastal and island regions are under-resourced in South Asia (Rajagopalan, 2008), with much of the current environmental discourse centering around agriculture, forests, land, and population demographics. The ocean has only come into public discourse in the region over the course of the last two decades (Balakrishnan, 1988). Consequently, islands like the Maldives, the Puttalam region of Sri Lanka, and the barrier islands of Kerala do not have the material sources for adaptation that would be necessary for the scale and scope of climate change impact over the course of time.

The town of Kollam (Quilon), once the transit home of Marco Polo, Niccola Conti, and Ludovico Varthema, among others, is a case in point. Today, the lagoon in the Ashtamudi Lake region is in deep trouble. Marine degradation, overfishing, and the anthropogenic impact of land and water use have produced a fragile ecosystem in need of more stringent environmental management (Kadekodi, 2004; Sengupta, 2001; Bhattacharya, 2001; Pereira, 2007). Corruption, illegal sand mining, and aggressive real estate development interests have depleted the delicate balance of island ecologies along the Malabar coast that Polo, Conti, and Varthema wrote extensively about (Venkataraman, 2005). Coastal flooding, sinking islands, and storm surge communities that are devastated by monster waves populate the region (Pande, 2014). The prevalent assumption that vulnerable communities along the lagoon fringe will draw on their resilience in order to continue living in low-lying barrier island environments is one underlying thread weaving this precarious intercoastal region into a lagoon ecology. There is plenty of creativity there, but the ecological catastrophe of waters rising in this delicate archipelagic region of South India is largely a tragedy waiting to happen (Joseph, 2019b).

The global North has made little attempt to really address the rising water levels and climate change issues affecting places like Quilon and its Ashtamudi Lake (Sallée, 2018). International climate mitigation agreements in the global North remain sadly inadequate to address the environmental challenges faced by cities like Venice and Quilon. As a climate concerned performance ensemble, we do not know what kinds of performances might be (im)possible in the near future. The new choreographies Harmattan is yet to design will continue to grapple with vulnerability and resilience, haunting and loss. We are now in a hiatus, waiting, writing, researching, and assembling ideas for upcoming interventions. For how long will physically distanced digital formats remain the norm, especially once vaccines are widely available? We may have to find ingenious ways to work across hybrid formats, welcoming technological platforms alongside live, outdoors, distanced modes of public gathering. After all, what the coronavirus crisis brought is nothing new but rather a halo of clarity on what has always been: the porosity and fragility of our bodies, our living condition of constant exposure to death and disease, the limitations of Western(ized) medical and scientific knowledge, the immense social disparities in access to healthcare and clean air, and how climate change and environmental devastation carry very real, concrete consequences right now, rather than hanging suspended in a future dystopian fantasy, waiting to unfold.

Interlude

Acqua Alta, Devil's Bridge, Venice, Italy (2014)

It is early morning. The island of Torcello lies in the distance, a 50-minute boat ride from San Marco. Four performers, the Venetians Tommy Girardi and Silvia Giudicci alongside Harmattan performers Varino and Bugge, take the old ferry to Torcello. The Venetian lagoon is dreamlike. Mist clings to the contours of the islands around the lagoon. The performers are headed for the oldest bridge in Venice, the Devil's Bridge or Ponte del Diavolo, a generic term for several rudimentary stone bridges built in Europe during the medieval period. It is a small stone bridge from the eleventh century with no handrails, and utterly mesmerizing in its minimalist design. Truly a place for *el diavolo* to come forth, summoned by a trick or spell. Girardi is a blues folk musician and plays a haunting composition filled with color and vibrancy. The four performers begin the slow walk across the historic bridge marking a journey in time from the eleventh-century Venetian mudflats to the high rising water levels of our time. Their gradual progression across the stone bridge into the forest beyond is breathtaking in the light of the rising sun. For Joseph, this performance is the culmination of a series of postcolonial returns. It was rumored that the Venetian Marco Polo had visited Malabar in the twelfth century. Before this, the Romans had traded with the Malabar from the first century CE. Thus a return to the Venetian lagoon was a navigational journey that made sense both historically and environmentally, in line with Harmattan Theater's commitment to using performance to draw attention to rising sea levels and climate change around the world. The Venetians have dealt with the challenge of storm surge for quite a long time now and have managed to accept its inescapable damage, yet remain flexible and proactive in the face of ensuing calamity. This performance was a marking both of their vulnerability and remarkable resilience. The phenomenon of *acqua alta*, or seasonal flooding due to higher than usual water levels, is an urgent problem in Venice today.

No longer merely confined to the winter months, it regularly inundates the city, creating a devastating ecological bind for the venerable archipelago: how to manage the flow of the Adriatic into the Venetian lagoon, how to cope with rising ocean levels, how to adapt a transformed human-made landscape of mudflats into a vision for the future?

Aqua Alta is a performative enactment of what we can learn from Venice. Like Venice, Cochin and New York are lagoon cities constructed out of barrier islands with furiously rising watermarks. The dredging of the channel around Venice to accommodate tourist ships has severely aggravated flooding across the islands, a familiar condition in and around Cochin and New York as well. The urbanist Edoardo Salzano (2014) observes that the engineering challenge posed by the Venice lagoon revolves around treating the lagoon as a regular body of water, and addressing it as a vulnerable and complex ecosystem increasingly out of balance because of voracious human use. Giant cruise ships and container ships are key threats to the delicate lagoon marine system. In Venice, Salzano bemoans, the terrible fate of the sinking city will only shift when humans reevaluate their relationship to nature and technology and find a way to reconcile conservation and tourism. Livability in Venice requires a relearning of what it means to inhabit a city of stone and water, as Salzano observes in *The Lagoon of Venice* (2014). This, alas, is not the direction Venice is currently headed. Harmattan's performance of *Acqua Alta* is the mournful recognition of the slow relearning humans need to undertake to coexist with cities bound by water, much like *Sea Dike* in Amsterdam and *Mar Português* in Lisbon.

For documentation of *Acqua Alta* (2014), please visit harmattantheater.com.

3 Harmattan Theater as Oceanic Praxis

Why Water Matters to Performance

May Joseph

New York City is an archipelago. Little islands dot its formation. The city rises along the edges of the East River, the Hudson River, and the Atlantic Ocean. This complex geography built around the rivers' edges demands a way of thinking about the city that is shaped by its water lines. Walking along the contours of Manhattan island in particular, one is immediately submerged in a landscape of forgotten streams, sunken ponds, and buried swamp lands: a hidden history of water flows.

Drifting through these traces of water flows year in, year out, I return to what architect Rayner Banham proposed as an architecture of ecologies. For Banham, Los Angeles was the iconic city of the early 1970s, forcing a fresh rethinking about urbanscape through the city's shoreline. Meandering along the invisible pathways of streams and canals of Manhattan, it seemed to me a good idea to dig deeper into Banham's invitation to pursue the hidden structures shaping how we imagine and inhabit space.

The catastrophe of September 11, 2001 brought home the intimate reality that islands are self-enclosed environments interdependent on modes of egress and entry. Islands are porous along their water's edge. Yet, their boundaries are closely regulated and monitored by public and private entities, such as the New York City Parks Services, the U.S Coast Guard, real estate interests, and the artificial barrier of expressways planned by Robert Moses in his creation of major parkways along the water's edge around the city (Gutfreund 2007: 86).

During the first few days after 9/11, there was no way to leave the city if you lived below Fourteenth Street. The trains were grounded, and cars were not permitted below Fourteenth Street. Through the ash of those early days, the structure of aquatopia began to take shape as an aesthetic practice. How does one engage with the hidden histories of island cities? Is there a way to physically impact the uncontrollable

DOI: 10.4324/9781003359906-3

extremities of waterfronts, allowing for investigation of habitations, of how we live and desire and hope on the edges of the built landscape?

A theory of water performance made its forceful appearance in catastrophe, as a growing awareness of the end of our water supply along with the dilemma of how rising oceans will impact the contours of New York City made it clear that the space of water is an urgent territory of engagement. Unpredictable, volatile, fluid, water as a site of human interaction raises intriguing possibilities for performance.

Water rituals exist in cultures around the world. Rituals of purification, of healing, of rebirth, and of appeasing the gods accompany the histories of rivers such as the Ganges, the Nile, the Tigris, and the Mekong. Along the "River that Flows Both Ways" (as the Hudson River was known by the Native Americans), any ritual relation at or to the water's edge had been severed by Robert Moses's indelible hand. The West Side Highway, the Brooklyn–Queens Expressway, FDR Drive, and the Henry Hudson Parkway sealed the waterfront away from the public during the middle decades of the twentieth century.

Automobiles got the best view of the city's waterfront as Moses had envisioned (Gutfreund 2007: 87–88). You would be hard pressed to imagine there was water a few blocks away, if you were not on the river's edge. If one was privileged with a view of the city's working waterfront, the chances were its noisy shipping industries made access to the actual river front a difficult and dangerous excursion to be avoided. Fetid industrial waste, rough neighborhoods, and abandoned waterfront properties created an aura of toxic absence around New York's vast shoreline.

9/11 forced a shift outward towards the water's edges, as the debris of the World Trade Center was carried onto barges moored at Battery Park City, and ferry services replaced train transportation between Manhattan and New Jersey. Michael Bloomberg's tenure as mayor further opened up the waterfront spaces of the city in a surge of voracious gentrification, creating new possibilities along with astronomic real estate realities. New York City's waterfront became simultaneously more accessible and less affordable. The edges of the river were opening up to the city through new public and quasi-private parks established along its riverfronts, which also displaced established residents and mixed-income communities along the once-dilapidated waterfronts of Manhattan and Brooklyn. It was in this unstable space of shifting urban morphologies that the Harmattan Theater took shape as a space for responding to our daily relationships with the waterfront of New York City.

Aquatopia

Harmattan Theater began as a local experiment with aquatopia in 2009—with water and space, grounded in the entrails of Manhattan's rocky shoreline. The idea catalyzed when the tsunami of 2004 devastated different continental shorelines across Asia and Africa. Mombasa was affected alongside Cochin and Aceh. This ecological interconnectedness was a terrifying scenario demanding new aesthetic priorities. Rising ocean waters seemed a dramatic *mise en scène* for an inquiry into water and its locationality in our lives.

Conversations with Victoria Marshall, an Australian landscape architect from Sydney who had created many public gardens and environmental public art projects around the world, led to a shared realization that Dutch navigation maps connected Batavia to India to Australia and New Amsterdam from the 1500s. This early maritime connectivity of water maps became, and still remains, the historic grounding for Harmattan's transnational projects.

Global climate change is connecting geographically diverse regions into one interconnected nervous global system. Aceh, Cochin, and Mombasa were affected by the same tsunami, despite their separation by large bodies of water, the Bay of Bengal and the Indian Ocean. New York City is learning from the tsunami that we must think ahead to a time when the earth's water line will have risen two feet in fifty years. This fact offers Harmattan's working imperative to probe the impact of water on local communities as well as sites around the world that were connected in earlier eras through navigation and colonial cartography to New York City. Harmattan's interest is to draw into conversation neglected and forgotten waterfront landscapes that are tied to New York's imaginary through history.

Palimpsests, Biodegradables, and Ghosts

Harmattan's first performance was on Governors Island in 2009, for the Figment Arts Festival. The year 2009 marked 400 years of New York's invention as a city. Drawing on its historic import as a date commemorating colonial conquest and the decimation of the Native Americans in the region, Harmattan Theater built a performance around the colonial discovery of the Hudson River, while also drawing attention to the Hudson's contemporary importance to New York City.

The Hudson River's claims to fame are many. First, as the conduit to the technological innovation of the Erie Canal, which led to the emergence of New York as a major port city. Some time ago, the

Hudson played an important role in the global environmental movement, as reflected in the much-fought-for Clean Water Act of 1972 and subsequent successes. More recently, proposals to exploit the Marcellus Shale deposit in upstate New York by a water-intensive and toxic process called hydrofracking has generated alarm; the controversial method could possibly contaminate the Hudson River and the water supply of 8 million New Yorkers.

This propelled Harmattan to look closely at the Hudson River as a performance site and source of stories. Hydrofracking along New York City's water reservoirs would be a catastrophic setback to 40 years of efforts to keep the Hudson clean. Under serious threat of unimaginable water contamination from plans to drill for gas along six counties in upstate New York, the Hudson River is, and remains, a precarious riverscape. Its fate is linked to ours as New Yorkers; and as concerned citizens, we wanted to add our voices to the chorus of environmental arguments opposed to an agenda dominated by big politics, and corporate and real estate interests.

Henry Hudson's Forgotten Maps brings the colonial history of the Hudson River into dialogue with its current vulnerability as a major watershed under threat of contamination. The performance begins with the little-known historic moment of Henry Hudson's arrival on the Hudson River. Based on the sea log of Robert Juet, first mate on Hudson's third voyage (and the first to Mannahatta), it unravels a forgotten story of colonial encounter between the Dutch and the Lenape Indians who resided on Notten Island, or what would come to be known as Governors Island.

Henry Hudson's Forgotten Maps was a movement piece highlighting the stormy waves and erased sandy shores of the greatly transformed original site of colonial occupation, Notten Island. The obscure fact that the first Dutch arrivals to the shores of Mannahatta had initially set up home for a year on the tiny Governor's Island seemed a critical piece of information to our project. The initial problematic of the piece was how had the waterline changed around Governors Island through the last four centuries of human use.

Work for *Forgotten Maps* began with a close study of the Dutch East India and Dutch West India Company trading maps. Mannahatta was only one tiny outpost in an elaborate system of trading ports and refueling stations along a vast maritime empire that connected the East Indies to the Americas. Traders in Batavia (modern Jakarta) and Kaap Staad (Cape Town) were connected to their Dutch compatriots in Mauritsstad (now Recife, Brazil) and New Amsterdam. Henry Hudson's world was overlaid with intersecting trade lanes of spices and slaves.

The performance Harmattan sought to produce had to reflect the material and mental landscapes shaping the fated encounter between Henry Hudson and the island of Mannahatta. I turned to the diary of Robert Juet, one of Henry Hudson's crew, for an idea of what that moment might have felt like. In Juet's account, there was trepidation among the travelers as they sighted Mannahatta. Awestruck when they first saw it, Hudson and his men did not land on the island. Drawing upon this historical anecdote, we created a script that drew from Dante Alighieri, Heinrich Heine, and Carnatic classical vocal traditions to create a sense of wonder and awe in Italian, Portuguese, German, and Sanskrit. These tonal variations reflected the myriad voices (over 30 nationalities) that already filled miniscule Mannahatta by the time Peter Stuyvesant became Director-General of New Amsterdam.

Governors Island, or Notten Island as the Dutch called it, has been expanded over time by infill, and its past is buried under macadam. To find the contours of the island's original coastline requires digging up maps that preceded the British colonial occupation of the island as a military base after the Dutch. This proved to be difficult, as the island had always been out of bounds to civilian habitation. However, the few images of Governors Island under colonial rule suggests a tiny area with a rugged shoreline (Glen 2006: 12). From these maps, along with the early soundings of Peter Minuit, Harmattan created a visual landscape along the eastern boundaries of the older part of the island, located on the northeast of Governors Island. To paint maps along the ground was a challenge as the New York City Parks and Recreation Office has strict rules about any sort of props or markings that alter the landscape of the island. Harmattan's sculptor Jose de Jesus and landscape architect Victoria Marshall concocted biodegradable paints from Hudson River water, egg whites, sugar, and milk and traced translucent lines of the glassy paint along the waterfront to create a gigantic backdrop for the Odissi and Kyogen-inspired movement of the performers.

Members of the company first danced the forgotten history of Governors Island, its sale for the price of two axe heads, some white beads, and a handful of nails to the rapacious Dutch who were there to stay rather than merely passing through, as was customary among the native peoples. The misunderstanding between the Dutch and the Lenape produced New Amsterdam: a stolen gift, a theft of good will. Using biodegradable materials for marking and etching the island's contours fittingly documents the fluctuating histories through which the island's pasts have been erased. Yet, despite Dutch claims to virgin territories awaiting colonial habitation, archaeological excavation unearths hunting arrows from a thousand years prior to the arrival of the Dutch, attesting to a long history of Native American presence (Wall and Cantwell 2001: 7).

A layered, palimpsestic trace of the many colonial histories and diverse populations that have occupied the landscapes of Governors Island and Mannahatta necessitates a more fluid, mutable account of the real story of the encounter. Such a narrative is impossible to fully capture, but its haunting reminders are constantly surfacing. To stage this impossibility, performers were positioned along a wide horizon, barely discernible to spectators. Performers simultaneously moved to the north and south, making it impossible to capture the entirety in a single glance. The viewer's gaze is drawn in opposing directions deliberately, forcing a realization that the desire to capture history as a singular event is an illusion of sight: actions are always lost to the viewer. Choreographed to move at an imperceptible pace, each fall of foot and each gesture of arm flowed in a stream of slowed-down movement. Each module of movement opens up the space between gesture and voice. Time is slowed down, and human attention is distilled to a dreamscape, moments flashing and receding like the crashing waves of Buttermilk Channel. The audience must wait, watch with patience, and immerse themselves in the process of visual exchange. What unfolds is a provisional urban ritual where all involved are part of the protracted transformation.

Henry Hudson's Forgotten Maps attracted an audience of willing participants in its slowdown. People wandered into the performance as it unfolded on the street, stopping to watch the entire 45 minute show. Encountering the performance in mid-flow, audiences watched curiously, loitered, and mostly sat in attentive meditation, allowing the physical slowing down of movement to draw them into a trance-like state. Spectators followed the liquid cartographic traces of a global history of maritime conquest: the now forgotten sea routes between New Amsterdam and ports such as Curaçao, Suriname, Cochin, and Macao in the Dutch colonies of the Caribbean, Asia, Australia, and Africa.

At the end of the performance, Henry Hudson sings a dirge in Portuguese recounting his first vision of Mannahatta as he wanders towards the water's edge, disappearing around the curve of the island. This finale was inspired by the tragic end of Henry Hudson, whose fourth and last voyage to North America ended in mutiny. Set afloat aboard a tiny vessel in the bay that would later bear his name with his little son and a handful of men by his defiant former crew, Hudson was never heard from again.

Virtuosity and Neoliberal Imperatives

Harmattan Theater's productions since 2009 investigate the boundaries between performance-as-product and the more amorphous realm of performance-without-end. Drawing upon Paulo Virno's concept of virtuosity as "activity-without-end-product," Harmattan's performances push the borders of materiality in different directions (2004). Most pertinent to Virno's theorizing of virtuosity, Harmattan's performances are frequently staged in sites of extreme physical isolation: remote edges of islands, the far end of a large pier in bitterly cold weather. The performances often have no audiences other than found ones. Actors perform in the rain or shrieking wind. The performer is reduced to an elemental form within the gigantic landscape of macadam, river flow, and the unforgiving elements of nature. Even the trace of the performance left on the ground (in the form of water from the Hudson River dribbled across the landscape) evaporates, leaving no record.

The idea for an environmental performance ensemble emerged against the backdrop of an onslaught of neoliberal initiatives in New York City. The evisceration of labor markets, shrinking of affordable housing, privatization of the arts, rampant reappropriation of waterfront properties for private residential developments, scaling back of welfare programs and support services for children, and a palpable collapse of middle-income opportunities across the city's small businesses and production sectors were signals that art had become unaffordable. Life in New York City had become very hard for the average family, with two-income households shrunk dramatically to single income situations, or worse, over the last decade.

In the arts, the scenario was even grimmer. While there are plenty of dreams to dream about as a city there are no spaces to dream in public as a theater company. Most public spaces are privatized, requiring permits and police notifications, and no one pays for non-utilitarian theater that is not commercially driven. For capital-intensive undertakings like theater, such an environment demands a return to older models of poor theater much practiced around the world.

The work of Jerzy Grotowski and Augusto Boal have long been models of creating theater across third-world stages, both within prosceniums and outside in the open spaces of the street. Indian street theater traditions of Jatra and Yakshagana, both ritual forms of outdoor performance drawing on found audiences, codified spatial uses, and non-actors, offer inspirations to undercut the grinding pressure to produce commercially successful pieces of entertainment that would hold the fickle attention of a demanding New York theater scene. To find examples of ensemble work that is self-sustaining in this city requires time travel: the Performance Group, Mabou Mines, La Mama, Wooster Group, and Richard Foreman's Ontological-Hysteric Theater remain, the last bastions of 1960s utopian possibility in New York.

On closer inspection, however, New York City reveals a panoply of small theater companies that produce challenging and experimental work on shoestring budgets, challenging the neoliberal logic that non-lucrative cultural labor must disappear with other inefficient economies such as illustration, print journalism, puppetry, or book making. As David Harvey notes: neoliberal imperatives invite ephemerality and provisional social arrangements (2005). This daunting reality follows the theater practitioner like a bleeding gut. There is no easy way out of the neoliberal world view. One can only offer partial interventions into what is the stranglehold of a profit-driven world economy. Producing free theater in abandoned and forgotten parts of the city is not a commercially viable enterprise. If one wanted to stage performances outdoors about the contamination of the Hudson one was going to have to be creative about it. Particularly if one wanted to achieve an ensemble production on a non-capitalist scale, working with a cast of 15 performers.

Harmattan's working premise was to work with the very categories of neoliberal efficiency that prevents theater practitioners from doing theater: work with non-actors the way African directors such as Gillo Pontecorvo, Djibril Mambety, and Sembene Ousmane did in such creative ways; draw from the pool of expendable bodies, those bodies that are traditionally not as lucrative in the theater, such as older actors. This principle seemed a fitting model for a city of outmoded lives caught in the throes of ever changing neoliberal economic strategies throwing people out into the cold. Out of the vast assortment of non-virtuosic citizens emerged a company of New Yorkers whose central auditioning factor was their immersion in the landscape of the city.

As director of Harmattan Theater, I was looking for performers who wanted to dance on the waterfront in hail, snow, rain, or sunshine. This is hard and grueling work, with no payment at the end of it other than the sheer thrill of dreaming impossible dreams in a city you love. It seemed to me that if I could find 15 people who wanted to do this, it would be a good reason to get the permits, find the landscapes, and create the time to do just that: citizen theater come to life as acts of urban possibility in a time of little utopian imagining. In Virno's terms, our actions were gestures of virtuosity in a culture of end products.

What resulted is a snowballing effect of passersby wanting to join a theater company that deals with the Hudson River and the more physically challenging spaces of New York's waterfront. People out of work, in search of passion, young artists, doctors, engineers, poets, architects, sculptors, philosophers, these were the motley crew of performers inhabiting Harmattan Theater. What brought them together was the ecstatic pleasure of immersing themselves in the waterfronts of New York for the non-commodifiable experience of capturing an inchoate sensation on the edges of a difficult city. I could not afford to pay these performers. What I needed from them was intense periods of immersion in physical movement over a large distance. This requires endurance, deep resources of physical strength, and an intuitive understanding of the site-specific landscape out of which the performance emerges.

What emerged through Harmattan's work with New York's waterfront was the incompatibility between conservatory trained actors and ecological theater companies such as Harmattan. Actors trained in acting schools are already inducted into the neoliberal agenda of profits and celebrity status. Most professional actors are in search of fame, a notable résumé, a coterie of rigorously trained performers to try their skills against. An environmental theater requires a whole different set of skills of which long hours in harsh weather, and the possibility of having no audience, is a regular feature.

Harmattan Theater begins with a site and builds performances around that site. Performers are generated around the site's needs: more dancers here, more sculptors and illustrators there. It requires an ecological mindset, immersion into the landscape, the sort of deep ensemble immersion that comes out of more socialist models of theater practice characteristic of Indian environmental theater movements, Vsevolod Meyerhold's notion of biomechanics, and Jerzy Grotowski's experiments in poor theater. In this logic of demanding, slow movement work without promise of an audience, the reward solely lies in the perfect event created through ensemble immersion in landscape, in the meaningful excavation of the history of a particular site. In that moment, theater momentarily finds relief outside the neoliberal economies of success and capital, of efficiency and productivity, as there is nothing efficient in working with the harsh elements of the outdoors, for a theater free to the public. It is a project in dreaming beautiful images about urgent issues such as water contamination and forgotten colonial legacies.

Space/Time Compression and Strategies for Environmental Theater

In an era of compressed lifestyles and difficult commutes, the challenges of creating an environmental theatrical practice are numerous. The simple challenges of working with a cast of 15 people who are all volunteers is enormous. Harmattan performers live in New Jersey, Westchester, Queens, Connecticut, California. Using technologies of space/time compression such as Skype, I began to train actors long-distance. Many rehearsals were done through solo one-on-one video-conferencing. The dress rehearsals and dry runs tend to be performances in themselves, as they are public enactments. To that extent, the strategy Harmattan takes is that every public rehearsal is a performance. Every performance is a rehearsal. This is a dictum that Bertolt Brecht practiced in his dramaturgical work. For Brecht, it is the rehearsal that is the performance. The process of theater is the point of the art, rather than the product itself. In environmental theater, where all variables are unstable—the audience, the weather, the synchronicity of element—there is no perfect end product, only a series of processes, a series of rehearsals.

Cuban film theorist Julio Garcia Espinosa proposed a practice of lucid creativity in the essay "For an Imperfect Cinema" (2005). Espinosa sought a more participatory, irreverent, dynamic, and public engagement with the artistic process. He critiqued elitism, exclusivity, and highly virtuosic artistic processes and argued for a more open-ended, porous, "imperfect," and democratic cinema. This 1970s strategy of challenging the orthodoxies continues to be relevant today even as the digital and networked economy have evacuated notions of truth, space, and a perceivable political spectrum and, instead, proliferated a fascistic irreality alongside democratic change. Espinosa's call for a theater of poverty requiring the highest refinement, a rallying cry in the 1970s avant-garde theater I came of age into in India, is empowering for a methodology of climate change intervention. The imperative to listen to the climate of change, blur the boundaries between spectator and performer, and decenter the spaces between everyday life and art has never been more critical than in this era of climate precarity. Consequently, Harmattan Theater's philosophy has drawn on many influential practitioners of an imperfect theater of which Ngugi Wa Thiong'o and Safdhar Hashmi are some of the more famous proponents. For both Ngugi and Hashmi, the space of performance is a liminal crossroads between the formal and the random, between actors and the everyday, between the open spaces of the outdoors and the codified arrangements of social rituals. This liminality is critical to environmental theater. It is the degree of uncontrollability that performing amidst the elements introduces into the arrangement of movements.

Finally, an environmental theater is an ecstatic theater. It releases the body out of the mechanical state of high productivity driven by the concept of the product, and demands a meditation on process, on the detail of movement in its minutiae. Antonin Artaud's experiments with dwelling amongst the Tarahumara were (among other things) attempts to touch the environmental unpredictability of living in the elemental forces of nature. Artaud did not find his ecstatic theater among the Tarahumara but his journey into the elemental suggests a practice of the ecstatic that is intrinsically secular. Vulnerability to the elemental forces of nature outside the proscenium draw from the actor an intensification of gesture, a concentration of energy that is trance-like precisely because the scale of the environment is expansive. Moving against the horizon inaugurates a performative state that is non-capitalist. It is the result of non-utilitarian conditions: difficult destinations, harsh weather, long distances, slow movements, a pronounced enunciation of ephemerality through sound and gesture. Harmattan's pedagogy of movement, drawing upon techniques of Kalaripayatti, Kyogen, and Vipassana, pushes the human body into an altered sense of spatiality, forcing a performance of intensity to cohere through an ensemble of performers.

Interlude

Far Rockaway, Tribute Park, Beach 116, Rockaway Park, Queens (2013)

Hurricane Sandy wreaked havoc in New York. Far Rockaway was destroyed by the storm. The site of Tribute Park, a memorial to the over 300 first responders from Far Rockaway who perished on 9/11, was erased. Beach 116, the street on which it is located, was deluged from ocean to bay. Catastrophically affected by the storm surge, Harmattan Theater wanted to address the question: Can performance be a tool of engagement for storm mitigation preparedness in coastal areas?

A medieval harpist plays sounds of the sea. An Odissi dancer immerses herself in the ocean. Dancers weave across a landscape of stones and sea. On September 14, 2013, Harmattan created *Far Rockaway*, a site specific, free, public performance at Tribute Park exploring storm surge topography, storm mitigation efforts, and community engagement, as part of the storm recovery process under way in the devastated region. Marking the vulnerability of barrier islands, the performance foregrounds the provisionality of living by the sea. Performers dance along the water's edge, immersing themselves in the icy waves as a way of reclaiming the affected site. Harmattan's goal was to bring the public back to a destroyed waterfront, and reclaim its restorative possibilities now ravaged by rising waters. Beach 116 residents whose homes and businesses were affected by the storm, were deeply moved by the performance. This was a community doubly affected, first by 9/11, then by Hurricane Sandy. *Far Rockaway* touched a chord of beauty, rejuvenation, hope, and mourning that residing on Beach 116 entails.

Harmattan deploys environmental theater as well as participatory action research methods in creating site specific performances. Harnessing community commitment and local concerns, Harmattan created *Far Rockaway* out of interviews with local residents, incorporating them into the performance's production. Audiences interacted with the performers, walking through the site of the ritual, engaging with gestures and questions. One spectator stumbled upon the show right after Temple, it being Yom Kippur. He said the performance was deeply transformative, as he had lost three comrades in the burning towers and had come to remember them, this weekend of 9/11 commemorations. The performance was cathartic for him. Another viewer, a restaurateur on Beach 116 whose cafe was destroyed by eight feet of water, expressed solace at witnessing what he felt was an "Egyptian" ritual, reminiscent of his Egyptian childhood back in Alexandria along the Nile river.

For documentation of *Far Rockaway* (2013), please visit harmattantheater.com.

4 Terrestrial Becomings

Walking for Climate

Sofia Varino

We are born into a world we did not make. This is perhaps the inaugural political and philosophical moment of conscious awareness, as philosophers from Hannah Arendt to Anne O'Byrne have noted. That inaugural moment entails a reckoning and perhaps a creative flicker, a glimpse of how (at least some) things could be otherwise. Making the world anew is also an act of unmaking: inhabiting, occupying, embodying the world anew. Harmattan's gestures are thus not of the utopian kind; as a heterogeneous collective, we do not hope for a promise of futurity as an improved version of our current condition, or a nostalgic longing for a pre-colonial, pre-industrial, pre-modern past. We do, however, want to open the world into its present, which is that fissure, that never meeting of historical past, dense and heavy, and a future still to come.

Inaugurated in 2021, *Plantation—Prosperity and Nightmare* is a monument by Angolan artist Kiluanji Kia Henda memorializing the role that Portugal, and in particular the city of Lisbon, played in the Atlantic slave trade by enslaving millions of Africans between the sixteenth and nineteenth centuries. According to the project's website, *Plantation* is also a memorial to "the resistance of Africans against the oppression of slavery" and to the contribution of Black and African communities to "Portuguese economy, culture and society" over the centuries.[1] The monument, a conceptual installation conceived within the scope of a call for participatory projects by the Council of Lisbon, is structured as a "plantation" made of steel. The history of Western philosophy is also a history of how to draw substantial rhetorical power from anti-Black and anti-African ideologies, as evident in genocide and slavery as they are in philosophical and scientific racism. These ideologies are inseparable from the extraction logic that subjugates all life to its economic value, so that land becomes property, body becomes labor, life forms become specimens to be classified into

DOI: 10.4324/9781003359906-4

species. Therefore, genocide, slavery, racism, and colonialism are all relentlessly at work within the environmental devastation that intensified extraction, exploitation, and expulsion have demanded on earthly bodies and places.

Plantation is an invitation to inhabit the familiar riverside urban space of Campo das Cebolas (now Largo José Saramago, named after the Nobel Prize winning Portuguese novelist) differently. The plantation is not so much fictional as performative. There is an uncanny quality to the monument, as if the many historical plantations it evokes were reenacted in real time. Walking through the steel structure is meant to restrict movement while also enabling creative, dynamic ways of moving across space. There is something to the repetitive patterns of the steel structure that requires walking with a certain rhythm, and that rhythm has a beat. The monument marks Portugal's colonial past while also referencing the project of European modernity as fueled by genocidal violence and extractive logic. Henda makes explicit the importance of marking public spaces with the colonial histories of violence they carry. Not far from the site where *Plantation* is located, Harmattan Theater performed *Mar Português* in 2012 (see the following Interlude). At that time, a commemorative monument marking the Portuguese role in slavery and the Atlantic slave trade did not exist. The Cais das Colunas, where we performed, was the point of entry for ships arriving and departing for several centuries. Our ritual of mourning and haunting aimed to engage with the Spirit of Water, its capacity for lively activity, while reckoning with histories whose weight we cannot bear. They elude us, even as we try to recall their relentless brutality.

In this chapter, I consider walking as a polyvalent tool for crossing research and resistance. I begin with a liquid memoir of watery dwellings from my childhood and adolescence in Portugal to think about the work my walking has done in collaboration with the walking of others, in Harmattan's site-specific performances. I proceed from the Tagus River, the riverside and estuary containing all my early experiences of water-bound life, and then move on to the Atlantic ocean, the foam and waves and tides and storms, before I cross an ocean to arrive at a Chelsea Pier and start walking with Harmattan.

The Tagus River as Multispecies Assemblage

The Tagus River is a key geographical and historical reference point in colonial history dating back to the fourteenth century, when the first ships begin to leave Lisbon through the Tagus estuary, advancing towards the Atlantic and initiating a centuries-long colonial enterprise that left a legacy of irreparable harm, from the Atlantic slave trade routes to the genocide of human and nonhuman indigenous populations. Today, the Tagus is also a toxic, highly polluted river whose water levels have been steadily rising. The Tagus River's seasonal flooding is as much a part of my childhood memories, something that seemed as unavoidable and predictable as the coming of winter. Images of people leaving their homes and being rescued flooded the TV news regularly. During flood season, the Tagus swells up and bursts out, turning into aquatic zones what are usually lush rice, tomato, and melon fields. This fluvial environment becomes then a zone of surge and crisis emergency, affected by heavy rainfall and seasonal storms. My grandmother was rescued from the second floor of her house several times, lifting herself over a windowsill to hop on a rescue boat as the water levels kept rising. The story filled her with laughter and a certain delight—she'd been through worse things; this was not a big deal. Unexamined, the Tagus overflow seemed to belong to the order of natural things we must collectively accept and endure. However, it would be fairly easy to make contingency plans for the seasonal flooding and avoid the annual loss of lives, income, and belongings. In this chapter, I pay close attention to walking as an (auto)ethnographic research method and as an aesthetic strategy in Harmattan's oceanic praxis. In other words, what kind of political and conceptual work does walking allow us to do? Walking, alongside the range of movements it supports and requires, is a way for anyone joining a Harmattan performance to slow down while remaining in motion.

In 2010, I started to systematically deploy walking as an ecologically attuned aesthetic practice in Harmattan Theater's environmental performances, drawing on the ritualistic and meditative experiences that walking can facilitate. Harmattan's work is informed by genealogies of ecological thought animated by anticolonial, antiracist, queer, and feminist ecologies invested in environmental justice. Terrestrial performatives thus encompass a vast range of political and aesthetic practices that are centered on action and enactment in real time. A Fridays for Future march, a Greenpeace sit-in, and the historical protests against nuclear war since the 1960s in the UK, all harness performative strategies for their political interventions. Harmattan's terrestrial performatives are informed by these movements and other political protest practices, as the ensemble reenacts a flat, situated, embedded, terrestrial ontology in large-scale site-specific performances.

Walking is a superb participatory research method as well as a multivalent pedagogical tool, allowing for physical immersion in the lived environment and spontaneous social contact with fellow walkers, performers, participant spectators, and non/human dwellers. In fields ranging from design and architecture to ethnography and geography, walking plays a key role both as research method and object of study. Walking is of course also part of everyday sociality, intermingling labor, travel, leisure, fitness, tourism, protest, and political action, epitomized by iconic figures like the vagabond, the *flâneur*, the stroller, and the wanderer, who explore and understand their surroundings through the immersive experience of walking.

Our walking is thus also a ritual of creative mourning. Nina Lykke's concept of "vibrant death" (2021) encapsulates and resists the tendency in much new materialist thought to attribute agency, enlivenment, and liveliness to inanimate entities, devices, and spaces. While this is an important task, recuperating an awareness of the participation of material forces in all aspects of social life, it also limits our understanding of the material to a restricted field of activity. Death and life are entangled in all forms of worlding. Walking along the Hudson River, the Tagus River, the Singel Canal, the Venetian Lagoon, the Atlantic and the Indian Ocean, Harmattan performers have been marking and mourning with their bodies the deadly-lively liminal borderlands of a new earthly paradigm. Operating according to a logic of radical presence, rather than the (utopian/dystopian) anticipation of futurity or the attachment of historicity, performing in durational, site-specific formats enables a rekindling of deep, rich time, in communion with life worlds.

How does walking with Harmattan, then, encapsulate a politics and poetics of multispecies kinship as radical ecological intervention? What kinds of decolonial, queer displacements, in a spatial and temporal sense, have we been performing for over a decade?

The Art of Repair

One of my recurrent childhood dreams is of a tidal wave. Tidal dreams are fairly common, especially for people who grew up or live sea bound. My friend Hinemoana Baker, a Māori poet whose name means "Sea", tells me of recurring nightmares about tidal waves she has had all her life in Aotearoa and since coming to live in Berlin several years ago. The city of Lisbon, as a focal point of seismic activity, is especially vulnerable to earthquakes and tidal waves, with the 1755 Lisbon earthquake, followed by subsequent fires and a tsunami, marking a radical shift in the city's history and architecture.

Growing up in Lisbon, I had a view of the distant sea from our fourth-floor apartment high rise up on a hill in the over-populated urban chaos of Queluz-Massamá, still the second most densely populated region in Europe, second only to Amsterdam. We often used our family set of binoculars to closely inspect the Atlantic Ocean in front of us: my mother and I took turns peering into the sliver of pale green, harsh blue, silver gray. *That* was the sea. A 20-minute car ride would take us to it, except for the traffic, which turned the ride into a two-hour ordeal. The sea was a constant presence. I found the ocean terrifying, as did my grandmother.

In *Undrowned* (2020), poet and scholar Alexis Pauline Gumbs proposes undrowning as an "emergent strategy," a concept from the title of the book series by AK Press publishing new forms of activism and social action that includes Gumbs's contribution to doing ecologies otherwise. Devotional, inventive, empathetic, and solidary, *Undrowned* proposes a poetic praxis in Gumbs's inimitable style, delineating strategies for undrowning in chapters on listening, breathing, resting, collaborating, going deep, staying Black, and slowing down. In *Braiding Sweetgrass* (2013), botanist Robin Wall Kimmerer observes that "[w]e need acts of restoration, not only for polluted waters and degraded lands, but also for our relationship to the world." Kimmerer advocates for the importance of learning the language of animacy and how non/human world forms of relationality are crucial for a radical ecological praxis that is transformative, rather than replicating the very systems of extraction and exploitation that continue to cause so much irreparable violence and harm. This kind of (un)learning begins with a listening (Gumbs, 2020) that disorders a humanist, rationalist, supremacist model of human exceptionalism centered on the capacity to "reason," using (human) language and exercising political agency.

For a radical climate intervention, data collection and empirical analysis are crucial but not enough. Activism aimed at changing policies and regulations is crucial but not enough. Renewable energy sources, green technologies, and climate mitigation programs are all crucial, but not enough. The decimation across all spheres of the Anthropocene (or the Necrocene, to build on Achille Mmembe's concept of necropolitics (2019) at work in the Anthropocene), demands a reckoning with all that's been lost and is continuously being lost. Restoration and renewal are needed just as much as acts of creative mourning and radical, slow, steady repair. Walking with Harmattan entails an oceanic praxis of witnessing, mourning, protesting, and documenting the ongoing devastation caused by colonial violence and extractive capitalism in every site where we perform as a troupe.

In *Down to Earth* (2018), Bruno Latour proposes a terrestrial praxis to counter the still prevalent planetary model at the core of dominant versions of environmental science and politics shaping public policy today. What Latour terms "the new climate regime" consists of globally networked phenomena where actants participate on an embodied, horizontal, situated, relational scale. There has been an underlying tension in Anthropocenic thought between what Bruno Latour calls a "terrestrial" (grounded, horizontal, flat, earth-bound) and a "planetary" (cosmic, vertical, spheric, outer-space-bound) view of climate change and environmental crises. To put it more bluntly, the political interests prevalent in the green washing ecological sensibilities of the white middle classes oppose and often obstruct the political demands inherent in an "environmentalism of the poor" (Nixon, 2011), as the brunt of industrialization has always been borne by the marginalized and the underprivileged. Terrestrial approaches, then, can be understood as grounded in environmental justice movements aligned with decolonial, anti-racist, Black, queer, feminist epistemologies invested in actions of repair and collective healing by denouncing and resisting the environmental damage perpetuated over centuries of extractive colonialism. Environmental justice movements evidence this tendency in their concern with the living conditions of bodies and communities and the kinds of harm they endure in specific sites and regions. Planetary approaches tend to overlook material and geopolitical situatedness, preferring to study and document environmental issues from a universalized, disembodied, exoplanetary perspective, proposing technological solutions that often rely on the very extractive industries that have massively contributed to climate change. A planetary approach to the Anthropocene cannot help but remain invested in maintaining the relative comforts of white heterosexual middle-class lives in over-developed wealthy nations at the cost of inflicting deadly violence on the vast majority of (non/human) earth-dwellers. On the other hand, a terrestrial approach seeks to account for the embodied earth archives bearing witness and giving evidence of the centuries of harm inflicted upon living bodies and places. The terrestrial is a way of doing history that is not content with a slim chronological linearity stopping at the present. The terrestrial has a knack for digging deeper and producing thick, sticky accounts of the past that naggingly interrupt the present and latch on to a haunted future. A terrestrial praxis can stand the unbearable weight of history's bones, piled up high and buried all the way down below. It is a retrograde, regressive mode of ecological thinking and doing.

The Deep Body

Both climate science and politics are saturated with culturally mediated fantasies about site, location, place, and space. In Lisa Messeri's ethnographic study *Placing Outer Space* (2016), the author shows how scientists turn outer planets into actual concrete (and in some cases potentially inhabitable) places. Messeri's study is useful for considering how the "earth archives" (Yusoff, 2018) of planet Earth are also open to this kind of imaginative (re)mapping and (re)worlding. Moving from a disembodied, spheric, planetary logic to an embedded, flat, terrestrial one requires defamiliarizing modes of immersive, sensory engagement with the lived environment. Such defamiliarization provokes a renewed aesthetic attunement to one's surroundings through estrangement, bewilderment, and wonder. Generating and compiling data about a given site through what I am calling here a terrestrial performativity does not rely on meta-datasets or the predictive logic of anticipation and futurity at work in algorithmic thinking. Rather, it is a method invested in gathering a different order of information that is not necessarily reducible to more dominant data genres such as numbers, graphs, diagrams, or what has been reductively understood to constitute a valid scientific fact. The qualitative methods of ethnographic work, including thick description and participatory observation, are closer to a terrestrial performative method. An expansion of sensing capacities allows each performer to observe both the animacy and the decay around them. This expansion of capacities expands far beyond real time or immersion in the lived environment of any given site. Expanding one's sensory arsenal enhances sensory, perceptive, and cognitive faculties, refueling social energies and reconfiguring political commitments.

Drawing on a distinction between direct action and what the author calls "indirect action," Lisa Diedrich (2016) analyzes divergent tendencies in types of health activism that preceded and later supported AIDS in the USA, including the Black Panthers and other Black and civil rights movements, the women's movement, anti-psychiatry interventions, lesbian and gay activism, environmentalism, and various strands of anti-colonial and anti-racist resistance from the 1960s onwards. For Diedrich, *indirection* encapsulates a necessary (dis)orientation towards perhaps less legible, applicable, or effective strategies. Indirection thus requires attending to what is not (yet) known and acknowledging that there is much we don't (yet) know that we do not know. Diedrich traces her conceptualization of indirection in relation to and in contrast with the direct action methods of AIDS activism from the 1980s onwards, in particularly ACT UP. But importantly, since her project aims to offer a historical account of how conceptual and political interventions since the 1960s have shaped AIDS activism, she traces her understanding of direction and indirection back to Rachel Carson's environmental approach to health. In a chapter from the classic of North American environmentalism *Silent Spring*, Carson considers the effect of pesticides, observing that

> [t]he new environmental health problems are multiple – created by radiation in all its forms, born of the never-ending stream of chemicals of which pesticides are a part, chemicals now pervading the world in which we live, acting upon us directly and indirectly, separately and collectively.
>
> (2002: 188)

The effects of climate change and environmental devastation can thus not be apprehended (documented, measured, explained) in a linear causal narrative precisely because they act upon human and nonhuman life forms *both* in direct and indirect ways. Radical ecology requires a capacity to see and act in transversal, askew, oblique, tangential, queer ways. Indirection as method enables a transversal engagement across disciplinary and political locations, encompassing activist modes that are often less immediately legible, but that are politically aligned with direct action strategies by supporting and expanding their effects. For example, Harmattan's large-scale, site-specific projects are informed by climate science data and climate change mitigation policies as well as by the geographies and histories contained in each of the sites where we deploy our terrestrial performatives.

Indirection might also provide a method for occupying and moving across space. Rather than being directed towards an object or destination, to be indirectly positioned may reveal novel, queer orientations and unexpected trajectories. Harmattan's choreographies have been developed by May Joseph as exercises in sensing and witnessing space through collective movement. As an ensemble, we move through space in tandem, sometimes aligned and other times as a constellation. Indirection becomes a method for sticking to the path while going another way. Indirection, then, is one way of performatively moving that emphasizes regularity and repetition, rather than progress and discovery. Whether moving in alignment or disarray, Harmattan performances are a becoming-terrestrial, unfolding their trajectories indirectly.

Considering the political potential of walking as a practice of physical motion and mobility also entails examining its multiple modalities beyond normative able-bodied walking to encompass assisted and prosthetic forms of movement. Part of what can make walking such a democratic practice is the relative accessibility of walking in public spaces for those who are able-bodied, and it is urgent to demand more accessibility for those with reduced mobility and special needs. Walking, especially in urban areas, is very different depending on whether one is able-bodied or differently dis/abled, presenting specific challenges when navigating spaces that are, by and large, designed for normatively able, "healthy" (and often male) bodies to move through. Unfortunately, urban planning often fails to support the needs of those with disabling conditions, from mobility to neurodivergence, especially considering how street traffic and mass transit systems are designed and structured in large, crowded cities. Modern societies are decidedly ableist in that there are often unacknowledged normative expectations of physical strength, speed, autonomy, and sensory and cognitive responses, thus restricting access and participation. This makes it unnecessarily difficult for the disabled, the elderly, the neurodivergent, and the chronically ill to have access to public spaces, whether it's to participate in a protest, enjoy a city tour, or join a walking performance. Public art has a duty to intervene and advocate for improved infrastructures and urban redesign to accommodate special needs, including more space, safety, and awareness for walking and moving in its various technologically assisted forms.

Walking shares a rich history with religious rituals like pilgrimages and processions or meditative practices like the Buddhist *Kinhin*. Historically, it has proven to be one of the most readily available strategies for large numbers of people to protest and demonstrate, as we have recently witnessed with the Black Lives Matter, Fridays for Future, and Women's Marches worldwide. Whether on a Sunday afternoon or at the end of a long workday, heterogeneous groups can come together and march through a bustling city artery or fill up an international airport, voicing their outrage and expressing their dissent across parks and plazas. In our digitally mediated and politically volatile times, walking remains one of the best tools we have at our disposal to stay connected while socially distancing and keeping in touch with those around us. Although it may seem like such a mundane part of our everyday lives that we often take it for granted, walking allows us to rediscover and engage with our habitual surroundings in novel ways, enabling close contact with human and nonhuman lives.

In a Harmattan performance, the synchronicity of our movements demands an adjustment to the slow pace of the walk, planned and rehearsed, as well as attunement to the person immediately ahead of us and whoever else is ahead of them. We cannot rush and we cannot delay. The music sets the pace of alterity that intensifies the walk and allows us to immerse ourselves in our habitual or novel environment while paying attention to a much broader range of sensory impressions: sounds, shapes, colors, scents, temperatures, textures. When participating as a performer or spectator in a Harmattan project, there is *always* time for rest, for respite, for replenishment, whether pausing to meander and to observe one's whereabouts or to listen to a performer recite Dante, Walt Whitman, or Fernando Pessoa. There is time to be *with* water and to *become* water, (re)encountering its flow, steady as a pulse, feet firm on the ground, eyes wide open, taking in the salty air. The balancing act of the slow walk meditation demands enhancing one's senses, really feeling the soles of our feet as they bounce against sand or concrete, grass or rock, gravel or cobblestone. The slippery surface of the stones smoothed by the river Tagus and flooded daily by its tides was challenging for Diana Niepce's sea sprite dance, my slow walk, and Marcelo's extended dungaree performance. Keeping our bodies grounded, balanced, supple, and upright at all times demanded constant focus, forcing us to move through that familiar site in new ways. The long duration of the piece and the minimalist choreography incited a kind of ecological intimacy with our surroundings over the many hours of filming and performing.

Walking Art

Artists like Marina Abramovic and Vito Acconci have created historical performances where the practice of walking plays a central role. Acconci did his influential *Walking Piece* in 1969, following a random stranger every day for a month through the streets of New York. Nearly 20 years later, Abramovic and Ulay performed their iconic *The Lovers—The Great Wall Walk* in 1988, in which they covered the distance of the Great Wall of China, walking from opposite directions and meeting somewhere in the middle, and then parting to never meet again (they did see each other again when Ulay was a participant audience member at the MoMA Abramovic retrospective *The Artist is Present* in 2010). At the Marina Abramovic Institute of Performance Art (MAI), mindful walking is very much part of Abramovic's pedagogy and methodology as an artist and educator. Historical artworks centered on walking such as these have received widespread critical attention since the 1960s as experiments in everyday life (other such artistic experiments have included eating, sleeping, and sexuality), testing the limits of what would later become known as "relational aesthetics," a term originally proposed by Nicolas Bourriaud in 1998 to designate participatory, performative, and social structures in artistic practice, most often used in the context of conceptual art. Walking has a special place in the history of women's performance art up to today, with names like Sophie Calle, Michèle Bernstein (of the Situationists International), or Lottie Child with her Street Training. In the performance *It Takes 154,000 Breaths to Evacuate Boston*, artist-scholar Kanarinka (Catherine D'Ignazio) ran through the entire Boston evacuation system to measure it in breaths. Although the piece is centered on running, rather than walking per se, the work is very much a piece of walking art in that Kanarinka uses the specific of her body's physical mobility to mark and follow a specific trajectory through urban space. The artist has also used walking in other art projects oriented towards civic engagement, environmental activism, and critiquing US securitization and surveillance policies since 9/11.

Walking has of course a rich history in European modern art that can be traced back to the surrealists, the Dadaists, and later the Situationists with their *dérive*. In my own genealogies of walking practices, I tend to associate walking in Western contemporary art within the context of specific events like the Dadaists' performative excursions and tours or the surrealists' enthusiasm for dream-based and psychoanalytically informed nocturnal strolls in Paris during the 1920s. Walking art also shares many of its aesthetic and methodological strategies with protest art, and outside of an explicit art context it is of course an essential tool for activism, including marches, riots, and public assemblies. In protest art and in social justice movements, walking has been a historically crucial strategy for getting large numbers of people to come together and move together, exchanging energy and information, fluid enough to enable movement while allowing for pace, rhythm, and political networks to emerge, holding a multitude of bodies together. This has been evident from historical movements like the US civil rights movement or the May 1968 protests in Paris, to more recently the Black Lives Matter multi-city protests around the globe throughout 2020–21, the Fridays for Future marches, or the Women's Marches in 2017. In contemporary art, walking practices are often coupled with performance and duration-based formats, especially in site-specific art, exploring lived environments, temporal structures, and the limits of what bodies can endure and experience as they move through space. Janet Cardiff's audio walks are exemplary works of walking art as well as sound art. A participatory format that has emerged in walking art is the audio walk, sometimes accompanied by podcast or app digital formats, which can facilitate immersive interactions among performers, spectators, and public spaces through the everyday activity of walking.

Walking for Climate

The climate ethnographies that Harmattan has been generating and collecting over the past decade are a repository of tales we cannot tell. Our encounters with place, with people, with the elements, with nonhuman actants, eventually form our theater of climate. Climates can be understood as intricate networks of performative activity where chemical compounds and living matter interact with meteorological conditions on a planetary scale that is also regional and locally bound. Geography is thus always simultaneously human and nonhuman, to the point that any previously held clear distinction between those two categories becomes irrelevant. In spite of its environmental commitments, Harmattan's work cannot be reduced to a public intervention nor to the idea of protesting climate change or demanding better environmental policies on a global, national, and regional level. What is it, then, that Harmattan proposes to do by using site-specific performative walks to participate in a broad range of measures and resources pushing for a radical environmental practice?

As a genre, endurance performance demands persistence and some degree of physical fitness coupled with mental strength. But most importantly, endurance performance requires tremendous attention to caring for one's body, tending to breath and balance, ensuring that the soles of feet are at all times making the right amount of surface contact, offering the right level of traction, pressure, weight. Endurance is thus less about surpassing one's limits than staying well within them, ensuring that sensory, cognitive, and bodily needs are met. Staying hydrated, avoiding sugar lows, managing sun exposure to avoid sunstroke, minimizing risks that do not add to the performance, such as wearing slippery footwear or inadequate clothing. Adjusting to heat and cold.

Duration-based formats emphasize the formal element of temporality. Walking as a form of protest has of course a long history, as does walking as a research method: histories of walking for leisure; as part of scientific, artistic, or religious pursui; walking as sport or fitness regime; or walking as social act, as a form of communal or political intervention. Mindful walking queers and decolonizers of space. Our practice is cumulative: through an accumulation of gestures, an aesthetic sensory experience for the performer and for the audience emerges.

Walking remains.

Note

1 For more information, see www.memorialescravatura.com/english (last accessed March 14, 2022).

Interlude

Mar Português, Cais das Colunas, Terreiro do Paço, Lisbon (2012)

A dancer with a didgeridoo weaves his way across the medieval flagstones of Terreiro do Paço. He holds the audience captive with the deep wind sound, filling the river air with mournful melancholia. At the water's edge, a nymph-like dancer in white, nearly imperceptible from afar, rolls with the Tagus River's restless motion. On the steps, dressed in stark black, the apparition of da Gama's daughter floats down the historical steps of the colonial site. The three dance a macabre sequence, trancelike, deathlike, merging with the sand, stone, and the river's rising tide.

The two-hour-long performance of *Mar Português* was inspired by the poetry collection *Mensagem* (1934) by Portuguese modernist poet Fernando Pessoa, and was performed at the site where Vasco da Gama's ships arrived after the first European sea expedition to India, laden with slaves, silk, and spices. Pessoa's articulation of "Mar Português" ("The Portuguese Sea") from *Mensagem* is a metaphoric, cartographic, and imaginary account of the Portuguese sea voyages. Born in Portugal and raised in Durban, South Africa, Pessoa returned to Lisbon in 1905, and this trajectory informs his interrogative and melancholic history of the Portuguese maritime past. In "Mar Português," a metaphysical oceanic space is laid out with coordinates in its colonial imaginary. The vast oceans of the Portuguese seafaring economy become unified into a singular overarching idea of an empire of the sea, which at its zenith was achieved by the Portuguese maritime empire. Pessoa's poem elevates the notion of a "Portuguese" sea as a political space of imperial might as well as a repository of colonial erasure. It both encapsulates a nationalist sentiment and etches the burdens of an oceanic metaphor rooted in slavery, pillage, and colonial conquest.

Inhabiting this liminal space invoked by Pessoa, Harmattan's production of *Mar Português* evolved as a collaborative postcolonial encounter between the past and the present, India and Portugal. Approaching Pessoa's poem "Mar Português" from the perspective of one of Portugal's former colonies opens up the thick history of the sea as a diasporic site of collective fictional and historical narratives. The concept for the performance emerged as Joseph was reading Pessoa's poem while staying in Vasco da Gama's bedroom as he lay dying in Cochin. Captivated by the poem's translation into English

> ("O salty sea, so much of whose salt/Is Portugal's tears! / … Was it worth doing? Everything's worth doing / When the soul is not small. / … Whoever would go beyond the Cape / Must go beyond sorrow"),

Joseph conceived of *Mar Português* as a feverish hallucination tormenting da Gama's mind as he trembles in delirium from a deadly tropical fever. The performance draws on the historical account of da Gama's final journey as Portuguese Viceroy to India. It was a tumultuous journey, involving notorious brutality on da Gama's part as he colonized East Africa and sought to consolidate authority over the Malabar region. At the height of his colonial maritime successes and infamously reputed for his acts of violence on the indigenous populations along the Malabar, da Gama was consumed by a devastating tropical illness and died in Cochin.

Harmattan Theater's *Mar Português* is a staging of the moment when da Gama is too ill to make the long journey back to Lisbon and dreams of the Tagus River, his only daughter, and the city of Lisbon in his final days. The performance mnemonically excavates the terrible horrors of colonization inflicted by da Gama on Cochin, through choreography and immersion at the historic site of da Gama's ceremonial arrival in Lisbon after his conquest of the Malabar. *Mar Português*, the Portuguese Sea, is about the return of the oppressed through performance. Joseph, a descendant of those colonized by da Gama, makes the reverse journey to Lisbon from Cochin to collaborate with Portuguese performers Varino and Diana Bastos to create a performance of colliding histories, of violent hauntings, of forgotten memories.

Set against the backdrop of the Tagus River's relentless tide, and the Monument to the Discoveries (Padrão dos Descobrimentos) guarding the river's mouth, *Mar Português* is a paean to the ghosts of the colonial past and to the new histories continuously emerging from that violence. It was performed as the culmination of a choreographic trilogy, conceived and directed by Joseph, that began in Cochin with the performance of *When the Sea Rises* (2011), which was created on the ramparts of the fort at Vasco da Gama Square in Cochin, a key site where the Portuguese first pillaged and colonized India along the Malabar coast. The second performance, *Cabo de Tormentoso* (2012), was done in collaboration with the Cape Malay community of Cape Town at the site where da Gama landed for the first time in South Africa, a cove first called Cabo de Tormentoso (the Cape of Torment) for the danger and nautical difficulty of crossing it by sea. Thus began a performance ritual of creative purges, one which retraced da Gama's sea route to Lisbon as a cathartic contemporary inscription of the repressed past that shapes the contemporary urban landscape across the Malabar, including Fort Cochin in India, where da Gama spent his last days, and Cape Point in South Africa. The urban past of Fort Cochin has now been transformed into the historical heritage of the modern tourist city of Kochi. Out of violence modernity ensues, relentlessly—and that is the story *Mar Português* begins to tell. The Tagus river front currently functions as a precarious public space for leisure, sport, entertainment, tourism, and artistic activity. *Mar Português* throws open the conflicting narratives engulfing oceans and their places within individual and national histories. While the personal story of decolonizing the past was one thread that wove through the project, the more urgent connecting theme was the new shared ecological fates linking former colonial port cities. Lisbon, like Cochin, is a water-bound city located on an estuary at the mouth of the river and the sea. What was critical 500 years ago to Portuguese colonizers remains even more pressing now: the strategic import of living by the sea. *Mar Português* is a material marking of that shared history. The historical weight of colonization now converges with the burden of living in the era of the Anthropocene, bringing to the fore issues of ecological survival on a planetary scale. Ultimately, the performance reenacts the realization that the distant past has folded into the immediacy of a common present. *Mar Português* materialized the interconnection of Cochin and Lisbon, between the rising seas of the sinking Lisbon waterfront and the deluging coastline of the Malabar region via seafaring and trade connections.

For documentation of *Mar Português* (2012), please visit harmattantheater.com.

5 Anthropogenic Citizens, Environmental Agents

May Joseph

Climate change has brought us as a planetary society much closer to a Greek theatrical logic than ever before—an archipelagic mode of becoming. Former cornfields are transforming into wetlands. Cities whose low-lying regions were built over are transforming into island-like flood zones on mainlands. The ocean is once again a critical *mise en scène* in the production of performance, art, and ritual worldwide since the growing awareness of climate and seas rising. We can learn from the great theaters of Athens, Delphi, and Ephesus—which all face the Aegean sea. Walking through these performance spaces, one becomes aware of the design genius of these ritual spaces. The seated audience in all cases face the ocean. At the Theater of Dionysus in Athens, the audience faces the Aegean in the distant horizon, beyond the city of Piraeus. The unfolding dramatic story of an Oedipus or an Antigone would have unraveled against the azure blue twilight of the Aegean, reminding its theater audience of its fragile dependence on the sea and climate. At Ephesus, similarly, the striking orientation of a theater setting facing the sea—now filled with silt and reclaimed land—reminds one how ecologically minded the early theater designers were. Performance was deeply related to its environmental reality as an archipelagic knowledge-making project. Ephesus faces Attica and the Aegean, as it would have beckoned its once multi-islanded denizens arriving at the port of Selzuk.

This architectural orientation of performance towards the sea is an ancient one across the ruins of the Greek theaters from Epidaurus, Eleusis, and Corinth, to Delphi, Sounion, and Ephesus. These environmentally attuned great stone theaters leave a spectacular trace of their ecological logic which shaped theatrical traditions across the Aegean. The sea in classical Greek theater design was part of the *mise en scène* of the stage. The ocean was the larger backdrop against which Greek society of antiquity imagined and participated in their

DOI: 10.4324/9781003359906-5

theatrical rituals. High up on top of the Acropolis, at the temple of Athena of Nike and at the temple of Etheneaus—women devotees of Athena would gather for rituals. They fasted and ate specific food in preparation for their secret rituals. During the era of Pericles the ancient Panathenaic processions that emerged out of the cult of Athena involved walking slowly from the Greek Agora below to the top of the Acropolis to worship the Goddess Athena.[1] Archaeological remains at Eleusis along the Aegean Sea document walking rituals from Attica to Eleusis for secret ceremonies to Demeter. Traditions of walking from Athens all the way to Delphi to worship the Asklupios, the God of Medicine, for healing were also ritual performances of devotion.[2]

These slow walks of immersive engagements with the shorelines of Attica and the Peloponnese took many days and much endurance. They were pilgrimages of trance as much as they were performances of devoutness by women devotees to the cult of the Mother Goddess, both Athena and Demeter, who are worshipped at Eleusis.[3] Walking along the coast was both the logic of the land and the orientation of the bodily practice. To walk along the Aegean shoreline was the meditation in movement as much as it was the vehicle of trance through which devotees reached Eleusis over a period of three days or more.

Drawing from these ancient Delphic pagan traditions of engagement with the shorelines of the Aegean islands, Harmattan Theater began walking New York's shorelines as a walking practice. The idea was to draw on a forgotten historic tradition of walking the coastlines of New York's archipelago in acknowledgment of the city as a settler colonial site. Walking slowly along New York's buried rivers, forgotten coastlines, built-up bays, and erased streams was a decolonial technique, harking back to a universal tradition of walking through watery landscapes. The idea was to begin to articulate a set of kinesthetic practices that were urban, easily accessible to non-actors, and engaged with the particular archipelagic imaginary of New York's island ecologies in a manner sensitive to its decolonial histories. Walking along New York's shorelines both connects with the settler colonial histories of the Americas and harnesses the ancient tradition of walking the coastlines of archipelagoes around the world, such as the Native Americans of archipelagic New York, and the Greeks.

Harmattan Theater began walking island coastlines and shorelines in 2009 as a decolonial praxis with climate, with environment, with settler colonial history. Our goal was to awaken ourselves to the contemporary challenges of thinking through climate as climate activists who wish to draw attention to rising seas, which in 2009 was not in the public awareness. Traveling through Delphi, Ephesus, Eleusis, and Argos, Joseph began to understand that Harmattan Theater's experiments with the slow walk along shorelines was in a longer tradition of practices across the Aegean during the fourth century BC. The ocean, Joseph realized, had very much been part of an ancient daily practice of immersion with the environment and ecologies of island societies. The slow forgetting of the ocean's power over the last 200 years of land bound movements and urban developments shifted our knowledge of these more enduring sea bound performance traditions towards a more inward-looking set of practices.

In a city such as New York, whose theatrical life is defined by Broadway and Times Square, located along Seventh Avenue between 42nd and 47th streets, the ocean has been all but forgotten as a theatrical *mise en scène*. Performance in New York has been inward bound since the closing of the great opera house on Castle Clinton in the early 1800s. Hence, Harmattan Theater's aesthetic preoccupation to use the ocean that envelops New York's over 40 islands as its theatrical space seemed to draw on traditions quite different from those of Broadway's commercially driven, entertainment focused logics.

The Lenape inhabited the coastlines of their New York archipelago as food was plentiful in the form of oysters and fish along "the river that flows both ways" that would eventually be called the Hudson. They would have walked along the coast, hugging the shoreline for its plentiful sea life. This was an ecology that Harmattan wanted to draw upon in its search for an oceanic aesthetic of performance. To walk along the coast of the island city of New York is to live in the path of an ancient logic of immersion in the geology and topology of hilly islands. This oceanic logic, of landscapes facing each other, Brooklyn, Staten Island, Manhattan, Governors Island, this was a specific kind of *mise en scène* that could form the reason for many performative investigations of ocean and land interfaces. Harmattan Theater's deepening engagement with an oceanic praxis is an organic response to the hydrography and riverine ecology of New York City, ensconced between river and ocean. An oceanic practice that is estuarine in preoccupation. It is a set of inquiries that could by extension be applied to many other types of riverine and estuarine locations with very different outcomes. This was the emerging materiality that Harmattan Theater was beginning to develop and delve into the mud, stone, foreshore, sand, and seaweed of oceanic praxis.

An oceanic praxis is one that seeks an engagement with the ocean's materiality, its histories, its transformations, its forgotten shorelines. Much of the world consists of coastal societies that have built up their shorelines in a manner that cuts the ocean off from its dwellers. The ocean is a place of respite, to be consumed as a commodity of leisure in modern day usage. It is rarely engaged with as a place of memory making, an ecology to be cared for and preserved, a resource to be treated with daily attention. Along the New York shoreline, the alienation of coastal peoples from their coast is probably the most telling. Four hundred years of a built environment that looked inward towards the city of Manhattan and rejected its working waterfront has distanced the ocean from the city's daily imagination. An oceanic praxis of performance in New York is therefore a larger philosophical project to revive the city's forgotten coastal imaginary as a seafaring city deeply immersed in its ocean's wellness.

The Cultural Construction of Women as Environmental Citizens

In his book *Citizens: A Chronicle of the French Revolution*, Simon Schama outlines the process through which ideas of the citizen were constructed.[4] The theater in Schama's accounting was a powerful public arena of display and contestation in the emerging conceptions of who the citizen was in the eighteenth century. Performance and performativity in the public domain signaled what citizenship could mean to the post-revolutionary French subject. During Donald Trump's presidency, the materiality of the ocean and the notion of a citizen were dramatically merged through deregulation and the devolving discourse of uncertainty haunting the Trump regime. A *New York Times* article listed a hundred environmental rules that Trump had either revoked, dismantled, or was in the process of reversing.[5] One could arguably state that we are in an unprecedented time of radical undoing at present. According to the legal scholar Donald Hornstein , it has taken over 150 years to build a legal history through which environmental laws in the United States have been created. In three years, Donald Trump decimated several of the foundational pillars of governance and citizenship. The idea of the environmental citizen as agent of action emerged amidst the chaos of environmental deregulation during the Trump era.

In our work with site-specific performances created at ecologically critical geological sites, Harmattan Theater has prioritized the notion of cultivating environmental agents of all stripes, men, women, children, non-binary people, by raising the question of what it means to live by the sea. By creating moving choreographies embedded in coastal landscape, Harmattan's aesthetics have entailed an ethics of multicultural environmental citizenship. How might we cultivate environmental agents through the public discourse of the theater, of movement, of dance? What does it mean to be a caring dweller of New York, a concerned environmental agent living in a vulnerable shoreline of archipelagic complexity? How should we engage with the materiality of living by the sea, of being impacted by the ocean's health, its unseen but tangible contaminations and toxifications?

Inventing the environmental citizen who is conscious about their impact on their archipelagic environment has been a philosophical imperative in Harmattan's meandering performances. Raising the issue of vulnerability, porous boundaries between land and sea, and the challenge of how to engage with ocean security is one thematic narrative that has structured many of Harmattan's performances to date. The very notion of protecting the sea around New York and elsewhere was a highly contested project in Trump's era, and continues to inform (post)Trumpism and populist movements in its aftermath, percolating down to extreme right-wing and conservative politics. The erosion of hard-fought Clean Water Act laws have created a new era of endangered water ways and water sources around the New York estuary. In the case of New York, an oceanic performance must be a political project about water precarity. The two are deeply entwined in our time.[6]

Since Hurricane Sandy, New Yorkers have gradually become aware of the ways that anthropogenic influences alter the balance and rhythms of the river and sea around us. However, the continuing education and cultivation of an oceanic knowledge that explores the relationship between coastal communities as environmental agents and their ethical responsibilities towards their coastlines and water sources has not been widely undertaken eight years after Hurricane Sandy. The importance of the work Harmattan Theater engages with, bringing people to the coast to investigate, explore, immerse, and understand their waterfronts, is a cognitive and cultural necessity for a deeper environmental transformation. People need to engage with their waterfronts in tactile ways, walking, dancing, embedding themselves leisurely in the ecology of their locality to realize that what they have is a fragile ecosystem that needs to be protected. The coast around New York is not adequately a live presence for its residents, as urban planning during Robert Moses's era disconnected residents from their shoreline. Many residents are unaware of the extent to which their lives are tied to the air and flows of water around their archipelago. This disconnection from the water caused by the construction of major highways such as the Brooklyn Queens Expressway, the FDR, and West Side Highways further distorts the impact urban dwellers can harness in improving their immediate surroundings through political and civic engagements with their waterfronts.

To that end, Harmattan's performances have been exercises in cultivating urban dwellers and neighborhood residents into becoming environmental agents. Drawing people into a landscape they walk through but never dwell on. Making people wake up to the nuances of the topography and structure of the coastline they take for granted. Luring passersby to pause and enjoy, meditate and marvel, at the ecological beauty of their archipelagic home with its interface of ocean and land, fluidity and hardscape, is a way to awake people into a realization that their environment is an agent of possibility. Harmattan's movement work, structured as slow walks along coastal waterfronts, reminds people to pause and consider the precarity of their environment, through the process of movement and stillness.[7]

Performers of the Sea

"The sailor was long a marginal figure in a society that … feared the sea" writes Jaques Le Goff of the Middle Ages.[8] This observation rings true for New York City and its coastal environs, as the city's post-sea-faring maritime era has entirely abandoned its waterfront and beaches to neglect. The sailor, the longshoreman, the sea captain were coastal characters of the early twentieth century that had fallen out of the city's imaginary by mid-twentieth century. Hence, the human connection to New York's coastline disappeared, leaving the extensive shoreline a place without identifiable human sociality apart from the extensive illicit life the abandoned piers and waterfront offered the city.

As the city gradually began to retrieve its sense of its coastal ecologies, citizen actions like those of Harmattan Theater began to emerge across New York's landscape. There weren't many prior to Hurricane Sandy, but enough artistic work engaging with the city's transforming shorelines probed issues of land use, reclamation, inundation, sustainability, and the importance of New York's seawalls in managing its coastline.

Harmattan Theater emerged as an organic response to the absence of engagement with New York's ocean-bound geographies. The choreographies were created to draw underrepresented groups such as women, queer, and Black, Indigenous, and people of color communities, the disabled, and the elderly to the shoreline, as a celebration and awakening of the forgotten waterline in their lives. For many people participating in Harmattan's performances, the effect of being performers of the sea is a tactile experience. To walk along the coastline, to immerse in the mist and froth of the ocean through performance, allows for immersive sensations creating a new sensibility of the performer as environmental agent. The performer is, in a Fichtean sense, engaged in the process of active looking-at that which the performer intuits.[9] Johann Fichte speaks of: "a looking-out of myself out of myself: a carrying-out of myself out of myself by the only kind of acting which is proper to me, by looking. I am a living seeing. I see (consciousness), and see my seeing (that of which I'm conscious)."[10] It is this living seeing that the process of slow walking affects the participant and viewer of the Harmattan performance. "Illuminated, transparent, unobstructed, and penetrable space, the purest image of my knowledge, is not seen but intuited, and *in it my seeing itself* is intuited."[11] The phenomenology of seeing the coastline, of opening up the spaces between land and sea momentarily, takes place sensorially for the viewer, the participant, the dancer. This space Fichte describes as "sensation is itself an immediate consciousness: I *sense* my sensation. This in no way gives me any knowledge of being, but only the feeling of my own condition."[12]

The process of creating performers of the sea is an education of the ad hoc environmental agent—the citizen actor momentarily transformed through an experience of sensation into a political subject concerned about ecological change. It is this space of performativity between movement and perception that the possibility for social change can occur. In this space between Fichtean sensation and intuition, the possibility of environmental consciousness arises. It is this juncture between the accidental sensations of performance encountered along shorelines by passersby, and their resulting sensibilities of ecological consciousness, that marks the efficacious moments in a Harmattan performance by the sea.

COVID-19 and its catastrophic effects on New York's public health have exacerbated the fragility of coastal communities open to the porous borders of land, sea, and air. The maps of those communities most affected by COVID-19 are superimposed on maps of those communities most affected by storm surges.

Anthropogenic Accountability

Interestingly, much of medieval history addresses the peasant against the agricultural hinterlands as opposed to the fisherman along the volatile coasts of Europe. Indeed, the fear of the sea that Le Goff speaks of shapes much of medieval social power in Europe. "Physically, the land is the oldest of all archives" writes Armand Fremont.[13] Fremont's essay on the role of land and peasant history in the construction of nationalist sentiment is instructive in foregrounding the unspoken presence the sea has had in discourses of citizenship. If land is the oldest archive in most discourses of nation-state inventions, the interface of land and sea embodies the space of ambiguity, the junctures of the bounded and the open ended.[14]

This silencing of the sea that informed notions of cultural citizenship in medieval Europe appears to have shaped early American understandings about land and water. Certainly in the accounting of New York, the plethora of histories of the city grounded in its terra firma imaginaries fill the annals of documentation. Less pronounced or documented is the vast and diverse histories of New York's multicultural coastline populated with largely immigrant communities and disempowered residents occupying frequently low lying, cheaper, and more compromised real estate located on former marshes, wetlands, and mudflats.[15]

Notes

1 Arnold Toynbee, *The Greeks and Their Heritages*, Oxford: Oxford University Press, 1981: 61.
2 Jenifer Neils, *Goddess and Polis: The Panathenaic Festival in Ancient Athens* Princeton: Princeton University Press, 1992: 15.
3 Frank J. Frost, *Greek Society*, Boston and New York: Houghton Mifflin Company, 1997: 96.
4 Simon Schama, *Citizen: A Chronicle of the French Revolution*, New York: Knopf Random House, 1989.
5 Nadja Popovich, Livia Albeck-Ripka, and Kendra Pierre-Louis, "The Trump Administration is Reversing 100 Environmental Rules. Here's the Full List," *The New York Times*, July 15, 2020.
6 "President Issues Executive Order Revoking Federal Sustainability Plan," Sabin Center for Climate Change Law, Columbia Law School and Columbia Climate School, https://climate.law.columbia.edu/content/president-issues-executive-order-revoking-federal-sustainability-plan-0 (last accessed July 12, 2022).
7 Harmattan Theater performances since 2009 include *Christopher Street Performance*, *The River that Flows Both Ways*, *Confluence*, and *Far Rockaway*, among many others. See harmattantheater.com for more information and performance documentation, including full-length videos.
8 Jacques Le Goff, "Introduction," *Medieval Callings*, Chicago: University of Chicago Press, 1996.
9 Johann Gottlieb Fichte, "Knowledge" in *The Vocation of Man*, Indianapolis/Cambridge: Hackett Publishing Company, 1987: 51.
10 Ibid.
11 Ibid.
12 Ibid.
13 Armand Fremont, "The Land" in *Realms of Memory: The Construction of the French Past*, Ed. Pierre Nora, New York: Columbia University Press, 1997.
14 Hugo Grotius, *The Free Sea*, Indianapolis: Liberty Fund, 2012.
15 Andrew Lipman, *The Saltwater Frontier Indians and the Contest for the American Coast*, New Haven: Yale University Press, 2015.

Interlude

Sea Dike, Singel Canal, Amsterdam (2014)

Sea Dike was an experiment in environmental excavation as well as a historical journey into the irretrievable past. In *Sea Dike*, we wanted to focus on the intertwining history of urban planning that shaped Cochin's early history when the Dutch usurped the area from the Portuguese crown in 1661. Bringing their knowledge of sea walls and dike building to the watery canal landscape of the archipelagic Malabar region, the Dutch had a strong geographic impact, tangible in the family genealogies, architecture, and naming of the early streets of Cochin: Utrecht and Zeeland. Haunted by this early history, the *Sea Dike* project follows the maps of Dutch cartographer and watercolorist Johannes Vingboons back to Amsterdam, where his painting of Fort Cochin is on display at the Rijksmuseum. The irresistible urge to create a performance at the edge of the city where land meets water led us to the site of Amsterdam's old moat, now Singel Street.

Staged as a performative return of the repressed, the performance materialized an ecological fact: that the contemporary reality of sea levels rising is historically linked to the radically altered coastal ecologies of formerly colonized regions of the world, such as Dutch India, particularly along the Malabar coast. *Sea Dike* articulated through movement the forgotten connections between Amsterdam's aquapelagic history and Dutch colonization's environmental impact on societies outside Europe. When the Dutch wrestled Cochin and the Malabar region from Portuguese control, they initiated one of the largest slave trade economies along the Pacific Ocean. This migratory route became critical to the creation of *Sea Dike* as Joseph followed a family rumor that one of her own ancestors, possibly a great-great grandmother, had been Indonesian. The rumors that haunt Joseph's family are corroborated by the travel logs of the Dutch East India Company archived in the Hague, documenting that two ships, the *Erasmus* and the *Niewenhoven*, did indeed arrive in Quilon in 1662 from Batavia, now Jakarta. The Dutch traded in chattel slaves and carried women as sex slaves on their ships from Batavia across the China Seas to India, en route to Cape Town. They kept meticulous documentation of their official slave cargo, but little record of the informal population traveling as personal chattel of the officers. The rest is silence. One way to respond to the past is through performance as a form of critical reflection. For Joseph, the medieval moat of Amsterdam, where the Dutch East India Company spread its vast and powerful tentacles, creating a world in its likeness, is an evocative site for enacting these entangled personal and collective genealogies.

Varino walks alone from under the Torensluis bridge, crossing it in slow motion towards Singel Street on the other side of the canal. A second performer, Alana Free, emerges from under the Torensluis following Varino, and a third, Marit Bugge, follows Free. A fourth, fifth, and sixth performer continue to follow in a slow stylized walk, drawing on the Kalaripayattu performance tradition from Kerala as well as the tradition of Vipassana dance from Laos. Both techniques necessitate slow, immersive foot work where the performer channels movement into and through the feet. The performers initially cross the street unnoticed, but gradually a stillness descends upon the bridge. Pedestrians join the walkers, and the line begins to extend. School children. Tourists. The human link crossing the Singel is a fluid crossing from the medieval to the modern city, across the historical moat. Dutch, Portuguese, Malabari, Turkish, Italian, the participants in performance come from regions that were intertwined in the early maritime trade. It is a historic journey of cultural rearticulation, recognition, and recalibration—perhaps closer to the radical ethics of forgiveness that Jacques Derrida proposes in his essay "On Forgiveness": "Forgiveness is not, it *should not be*, normal, normative, normalizing. It *should* remain exceptional and extraordinary, in the face of the impossible; as if it interrupted the ordinary course of historical temporality" (2001: 32; emphasis in original). The line of human bodies joined in a movement-based performance across the Singel canal enacted perhaps one such moment of this radical, transformative, exceptional, and extraordinary interruption of an inescapable history.

The early medieval Zeedijk built by the Dutch that still stands in Amsterdam also inspired *Sea Dike*. Today the world is learning from the Dutch how to manage water inundation, how to allow the water in and work with it, rather than insist on futile attempts to resist its force. Ancient colonial port cities like Cochin and Quilon are still using Dutch infrastructure, including canals, in their ongoing reimagining of the city. Yet the violence of Dutch colonization remains engraved in the visage of the coastal people of the Malabar and is embedded in the traces of Dutch architectural heritage, a daily reminder of the same haunting history *Sea Dike* evokes.

For documentation of *Sea Dike* (2014), please visit harmattantheater.com.

6 Queering Climate

Ecologies of Historical Radiance

Sofia Varino

We were scheduled to do a talk and a performance at Devil's Bridge, Venice's oldest bridge, in May 2020. Our talk and the performance were meant to address Venice's aquatic histories and how climate change is jeopardizing its ecosystems. As the conference was canceled, the Covid pandemic came to occupy a central role in everyday life particularly for those of us living in high density urban areas. No longer an afterthought, climate change, deforestation, and habitat loss came to occupy our delicate "human" daily embodied existence wherever we were, finding a way into our collective corpus, and coming to inhabit many individual human bodies. This was also the time for the Black Lives Matters movement to further establish global networks beyond its solid foundation as a US movement, instigating and supporting racial justice movements on a transnational scale—a time of reckoning and collective action.

The coronavirus pandemic encapsulates several of the current crises of our times: ecological devastation, scientific uncertainty, biosecurity, healthcare. We do not know what kinds of performances might happen in the future, whether digital or socially (physically) distanced formats might become the only or the more viable option for the kind of work we want to continue doing, caring for the planet and for each other. We may have to find creative ways to work across hybrid formats, welcoming technologically mediated alongside live, distanced, and intimate modes of collective performativity. After all, what the coronavirus crisis brought upon us is nothing new but rather a halo of clarity on what has always been: the porosity and fragility of our bodies, our living condition of constant exposure to disease, the limitations of Western(ized) biomedical knowledge, the immense social disparities in access to healthcare and clean air, and how climate change and environmental devastation have concrete, immediate

DOI: 10.4324/9781003359906-6

consequences, rather than hanging suspended in a future dystopian fantasy. We live in the age of social distancing, perhaps more accurately described as *physical* distancing. Restrictions on social life have led urban dwellers to socialize in open, public spaces and resulted in creative uses of parks, plazas, and streets that are closely aligned with Harmattan aesthetics. The tools and strategies we deploy come to form an ethos of uncertainty aligned with experimentation and improvisation.

Becoming Radiant

In *Climate Trauma*, E. Ann Kaplan (2015) proposes the term "pre-traumatic stress disorder" for the range of future-oriented anxieties that so many now share in the face of drastic environmental changes. Both the fictional and the performative can function as ways of experiencing, of documenting, of somehow making sense of events that may not have yet happened but can nonetheless happen anytime. Or to account for what has happened many times over, seems immeasurable or imperceptible, beyond all comprehension. Geological imaginaries are one way of creatively encountering and delineating those interminable timelines that geological epochs summon, a method for visualizing and imaginatively materializing massive time scales into something concrete and observable. The Anthropocene is but a conceptual container, a word signaling a specific timeframe, mobilized to articulate intricate phenomena. Kathryn Yusoff's *A Billion Black Anthropocenes or None* (2018) resonates with Harmattan's work not only for its anticolonial anti-racist critique of the limitations of the "Anthropocene" epoch and the dominant geological thinking embedded in it, but also in its call for a critical geology attuned to the inhumanities of knowledge production and the labor demanded from those labeled "inhuman": non/human populations, places, substances. The history of colonialism is the history of inhuman labor, forcefully extracted from those deemed less than human.

Anthropocenic thinking styles tend to rely on modern scientific epistemologies that corroborate the foundational distinction between a rational, politically entangled "humanity" and a mass of indistinguishable "nonhuman" matter distributed among arbitrary categories like animal, vegetable, fungal, bacterial, or mineral. This amorphous mass teleologically foregrounds human exceptionalism—since modern "humans," that is white, male, able-bodied Europeans, have been capable of such feats of violent intervention on planetary ecosystems, that surely proves their exceptional status above and beyond nature. The corresponding narrative is that "natural" life, uncontainable, wild, volatile, is now taking its revenge on "us" for the hubris of human exceptionalism. And yet, it is the marginalized and underprivileged who are more exposed to the effects of climate change, pollution, and toxicity. Citizens of affluent nations may still believe that recycling and curbing consumption, alongside devising clever infrastructures for mitigating environmental crises, will ultimately keep their worlds intact, while the rest of the world perishes. For better and for worse, this will not be the case.

Instead of a future-oriented logic of anticipation, prediction, and prevention, terrestrial performatives are oriented towards inhabiting the wild, disorderly present, extending themselves viscerally into the deep geological past, exchanging the attachment to controlling future outcomes for a sense of radiance that can render the present as uncanny while facing the future as an indeterminate unfolding. Such radiance is less interested in cultivating certainty through repetition of what seems safe or familiar, the fantasy of a return to less troubled times. Radiance can be a method for reimagining relations and infrastructures in the lived moment, considering their contingency and interdependence. Catherine Chin mobilizes the concept of "historical radiance" (2017) to account for how the radical, unknowable difference of ancient cultures and events can summon a sense of alterity that is as unsettling as it is wondrous, shaking habitual thought patterns and disrupting expectations. In "Marvelous Things Heard," Catherine Chin (2017) proposes the concept-method of "historical radiance" as one way of doing historical work that can account for estrangement, highlighting the incomprehensible, marvelous, and puzzling aspects of life during various historical times, and how this estrangement can inform our encounters with the present moment. Historical radiance might then be described as a practice of estrangement and perplexity attuned to how excessive, uncanny, liminal elements function in historical processes and events, especially those so distant from the present that they acquire an otherworldly dimension. As a historian, Chin is specifically invoking the eerie otherness of ancient and premodern worlds whose logics and events seem strange, weird, fantastic in ways radically distinct from what we expect. A radiant historiography is important for Chin as a research method for how it enables encounters with the past that are based on empathy, solidarity, and acknowledgment of radical divergence. Harnessing Chin's method of historical radiance, Gumb's emergent strategies for undrowning, and Bennett's recommendations for how to foster enchantment in the modern world, I want to consider walking and other forms of meditative movement deployed by Harmattan as environmental radiance. Extending the past of premodernity and antiquity into a prehistorical, prehuman, geological past encompassing the vast expanses of time before the arrival of life, this radiance may also inform an ecological praxis less devoted to the enduring logics of maintenance, resilience, survival, or innovation. This may be an ecological praxis for those who realize they/we do not (yet) know how to stay alive in the volatile planetary or terrestrial paradigms of a new earthly regime. Remaining earthly may allow citizens of overdeveloped affluent nations to reimagine solidarity, vulnerability, kinship with all life forms and life worlds. As the privileged few lose their formerly dehumanized selves, a fiction fabricated from the resources harnessed via centuries of a violent extractivist logic of accumulation, returning to a common earthly condition is no longer a choice but an imperative.

Chin goes on to call for forms of environmental justice and climate interventions that are not only mindful of nonhuman as well as human lives, but that in fact blur given distinctions between these two contested categories. Environmental performance demands endurance and the willingness to occupy given and familiar public spaces anew, to render them strange and radiant, to inhabit their historical radiance with a stoic perseverance through the elemental. Instead of imbuing these sites with agentic and subjective capabilities modeled on human experience and oriented towards human use, our method as a collective has been rather to enter into the historical radiance, the vast geological temporal scales each site carries, the weight of historical events encompassing horror, violence, loss experienced by racialized, gendered, and classed bodies. Entering oceanic, elemental, plant, climate time means a kind of temporal attunement to another pace, another rhythm, another beat. Meditative slow walking enables a slowing down of vision, as the forms of the lived environment quicken and sharpen, the contours of bodies, stones, and trees soften, and bodies of water lose their soothing surface blue, acquiring texture and temperature and a million hues of green and gray, specks of gold and bright white, skies turn orange, wind suffocates breath and heat sears flesh. Chin points to the importance of assigning oneself a kind of "imaginative 'weightlessness'" whereby we de-imagine our experience as singular, isolated, coherent selves, requiring that we "encounter otherness in its full disturbing weight" (489).

The fact is that much of the environmental damage caused is not reversible—not within a temporal framework relevant for human life. Although climate change mitigation technologies and policies can counter to some degree the devastating effects of rising sea levels, the harm inflicted by fossil fuels will persist. Ultimately, the very reason to intervene is this very irreversibility, the effects of which are already deadly for so many. Irreparable, the damage done can nonetheless be managed differently by attending to the recurring violence and devastation taking place in real time—right here, right now. Caring for the human and nonhuman lives impacted by the sprawling life worlds means also changing everyday practices and acts as well as working to change policies at the local, national, and international scale. It means demanding funding for scientific research and technological development, as well as a massive redesigning of school and higher education curricula towards a recentering of all disciplines around ecological knowledge across the natural, computational, social sciences, arts, and humanities, collectively developing environmental faculties of awareness, attunement, sensing.

The Lure of the Social: Relational Aesthetics as Ecology

Harmattan's work is closer to landscape art than theater, closer to large-scale installation than performance art. Our site-specific performances are political, geographical, philosophical, aesthetic interventions on public space and the lived environment. We have sought to performatively engage with the colonial past, not to reenact it but to acknowledge its ghosts, its perpetual haunting, its inevitable, brutal hold on the lives we are able to live in the present. Performance then functions as an interruption; it affords a way to physically and sensorially occupy the present, both in a spatial and in a temporal sense. It provides an aesthetically and socially attuned method for multispecies collaboration in public space. It is a practice of recitation with critical distance.

In *Social Works: Performing Art, Supporting Publics* (2011), Shannon Jackson addresses again the issues of legibility and accessibility that have been routinely raised in debates about the function of art in contemporary society, perhaps most memorably by Claire Bishop. Considering contemporary art, Bishop argued for the political role of opacity and illegibility in the influential *October* article titled "Antagonism and Relational Aesthetics" (2004). This emphasis on participatory, durational, social, relational structures in contemporary art is informed by temporal structures in performance art and dance as well moving image, video, and multimedia art. These formats have (re) introduced the live and durational elements of what had been conventionally classified as the "performing arts" (dance, theater, music) in contradistinction to the "visual arts" (usually referring to ostensibly atemporal media like painting, sculpture, or photography). With this new focus on sociality and participation, contemporary art practices seek to confront, challenge, and escape the confines of the the gallery, the museum, and other institutional spaces as (post)colonial, neoliberal, heterosexist spaces of cultural reproduction. With a bold commitment to conceptual and immaterial art practices, many leading contemporary artists are thus far less invested in producing commodifiable art objects than in investigating aesthetic experience and artist production, pushing the limits of what art is and can do—philosophically, politically, formally. Within this relational paradigm, an aesthetic practice then becomes a social practice embedded in everyday life. The interruption a performance affords is a lived, embodied one situated in real time and space: time pauses, contracts, and expands; spaces flow, overflow, and become saturated.

In *Notes Toward a Performative Theory of Assembly*, Judith Butler (2015) advances a theory of collective action centered around the practice of performative physical encounters among ephemeral, temporary publics. Butler's emphasis on performativity, informed by poststructuralism, linguistics, and speech act theory, has since their foundational *Gender Trouble: Feminism and the subversion of Identity (1990)* shaped the fields of gender studies, queer theory, and performance theory. Ultimately, performativity mediates Harmattan's encounters with water. Becoming-water entails engaging with the performativity of water as object, water as praxis, water as philosophy. Water is an unsettling object of study. The terrestrial order of things requires imaginative engagements in real time, rather than the detached anticipation of projected futures. Aquatic performatives are always provisional hybrids, demanding quick adaptation and daring improvisation. Walking around with a sense of radiance demands a kind of presence that can tolerate and sustain some degree of turbulence. Much like the figure of the *flâneuse* or *flâneur*, Harmattan performers and spectators enact a kind of sited, situated, embodied encounter with the lived environment through their trajectories, drawing with the shape of their trajectory a mapping of turbulent, ephemeral landscapes.

Symbiotic Performatives

As an environmental performance ensemble, Harmattan is oriented towards the lived environment as the necessary condition for art, and indeed all living things, to emerge. Queer phenomenological and ecological methods mix and mingle. Along the lines of Bennett's political ecology of things in *Vibrant Matter* (2010), Harmattan prefers tales where the agency of matter and the visceral materiality of human lives takes precedence over the mastery of "human" action and reason. An amphibian, terra-aquatic agency is understood to encompass both activity and passivity, the capacity to act and respond as well as the conditions of being acted upon, and thus modified and shaped accordingly. The intersticial spaces among terrestrial, aquatic, and atmospheric life worlds, merging earth, water, air, require an amphibian logic. We generate greater environmental attunement through aesthetic and sensorial means: the fluid, open structure of our actions in real time, documented via the digital technologies of photography and video.

In the *Chronocorpus* blog we started in 2010, a heterogeneous mass of text, photographs, videos maps, and historical documents hang suspended in digital space. The seriality of the content, published over a long period of time, suggests our commitment to keep a steady flow of material streaming into the WordPress platform. We struggle to maintain it among our many professional and family responsibilities. Eventually, the blog becomes a dull obligation and soon after an incongruent archive, documenting a snippet of our work with Harmattan. Nonetheless, it is to the blog that I turn several times to double check dates and look for input to stimulate my memory. Although the official Harmattan website preserves most of our performances, the disjointed materials floating on the blog stimulate vivid reminiscence precisely because they are peripheral, erratic, random. Their nomadic presence captures something of Harmattan's oceanic praxis in its fluid spontaneity, its outbursts of activity, its willingness to seek temporarily found audiences and to linger in public space, perhaps unnoticed. There is something profoundly queer about this ethos of failure, as Jack Halberstam (2011) has observed, a queer failure that is more about joyous improvisation unconcerned with meeting preconceived expectations than with any lack of means to act or create. Considering energetic supply in social and aesthetic terms, an ethics of failure is also oriented towards lowering costs, economic and otherwise, investing less in the presentation of a finished product than on the immediacy of process, of embodied experience, of relationality in communion with others.

Spectatorship and Participation: Audiences, Publics, Communities

If our performances are mapping exercises in real time, there are radically queer ones, producing askew, disproportionate, sticky, muddy, wet maps. They are not useless maps though; they propose and contain a specific way of situating living bodies in relation to space. They are wide open maps. They are self-generative and rhythmic, visceral maps with a beat. They hold within them an affective landscape of violent histories and future longings. They contain a language for decoding spatial surroundings through the steady flow of slow walking and a meditative state lifting the schism between the right here, right now, and what has come before, what has been here for as long as water. The sheer energy of water, its capacity for force and movement, is what enables its transformation into hydraulic sources of energy. The kinds of becoming-water enabled by performing in a dilapidated pier in Manhattan or across an ancient Roman bridge in Venice.

Performing in a queer decolonial mode also entails the enactment of systems of care and support. When performers come together for rehearsals, blocking the performance and going over our lines and choreography, we are also interested in building community. The modalities of ephemeral relationality that performing in the open affords, using contested public space to create an aesthetic experience that is also a protest and a ritual, are also a moment of queer kinship. Performing with the living world and the elements in increasingly privatized urban spaces is not only a gesture of defiance but one that catalyzes a communal exchange. Found audiences are drawn to the spectacle of bodies circulating in unusual patterns across familiar or neglected spaces, but they are also invited to join the show, as they did to spectacular effect during *Sea Dike* (2014) in Amsterdam. A Harmattan performance incites queer, decolonial uses of space that are not oriented towards (re)production. Indirect, transversal, elliptic, incomplete, our choreographies agitate the habitual and propose a liminal morphology for mapping, moving, and being in/with one's surroundings. How large-scale, site-specific performance offers a methodology for ethnographic and historical research that is embodied, situated, relational, durational, a fluid methodology.

Walking around a city is an activity with its own sets of rules according to the geographies of the city, its human and nonhuman inhabitants, its infrastructure, its design, and architecture but also its pace, its tempo, the histories it reenacts daily. In New York City, negotiating encounters with strangers and with the living world, the built environment, the Manhattan grid, can be grueling. Sensory stimulation and a constant demand for cognitive engagement make Harmattan's invitation to slow down a particularly appealing one. In other urban hubs, taking time to consider how one moves across public space can feel just as luxurious. We are enabling improvised interactions with water-bound areas to draw attention to their activity, the ecosystems they hold, the histories they contain, the futures they may afford. Engaging with water as a living, shape-shifting, agentic substance as necessary for survival as it is threatening, rather than presuming its existence as a passive resource. Against a romanticization of Nature with a capital N, engaging aesthetically and ethically with its elements, landscapes, weather, organisms, in ways that are sensorial, physical, temporal, provisional.

The coronavirus pandemic has forced us to reckon with spatial, global, and collective modes of thinking, and reenergized our social creativity in how we move and participate in the public sphere. The pandemic can be understood both geographically and environmentally. Environmental crises bring to the fore and force us to confront in the present what may have seemed abstract, suspended, hanging in the air like viral particles amidst an assembly of toxic particles, fungal pores, pollen spores, bacterial populations.

Queering climate demands an encounter with radiance as strangeness and as alterity, with historical radiance that disrupts linear time and flat space, beyond utilitarian purposes. It demands a recognition of how the inhuman conditions of the present are contiguous with centuries old systems of extraction and expulsion, forcing migrations and generating deadly conditions for too many beings. It releases us from a jaded familiarity with how things are, throwing us into the throes of dreaming up what they might yet become. The weight of history distorts and disturbs. To reclaim a site in order to inhabit its strangeness, a way to insert engagements with the nonhuman, efforts to dehumanize, demands new forms of relationality. Rendered strange, sites acquire an ahuman dimension, seeming to move into a future after human extinction, a future replete with geological, mineral, bacterial, and possibly fungal or vegetal life. An ethics of radiance demands accountability for the enduring violence of the present and how the ghosts of the past haunt our daily attempts to stay alive.

Epilogues

Harmattan Wind—Climate Change Aesthetics and the Nonhuman

By May Joseph

The orange haze of California's fires has brought home the prescient news: climate change is here to stay. We are engulfed in the histories of the nonhuman. Water, fire, air, earth. The world is a theatrical spectacle gone mad. From eerie stillness during the first weeks of COVID, to the terrified migrations from the hills of Northern California, and the displaced migrants of India, to name a few, our species is gripped in a series of dramatic mass mobilizations connected to climate and species precarity. We are in an era where the theatrical is everywhere, and the stage is the world. Our script asks the urgent question: *What is to be done?*

As citizens of the world, we are engaged in a grand narrative of the planetary. The gigantomachy with their gargantuan powers of nature have been thrown out of balance by the extractive acts of human greed. Now we have run out of time to imagine the consequences, as they have arrived with little warning. As citizens of the world, we must engage, grapple, respond, immerse, translate. We have to cultivate a new language of creativity that incorporates the nonhuman as part of our coexistence. The work of Harmattan Theater has been an embrace of this challenge to capture the vast possibilities between ocean, wind, stone, humans, and sea histories. It has been Harmattan's goal to embed its performances in coastal and island landscapes and allow the weather, nature, and the attending elemental forms to structure the performance through movement and elemental sounds. An underlying motivation for Harmattan's work has been to excavate and unearth a vernacular for climate aesthetics through the juncture of human and nonhuman engagement, at the cusp of movement and macadam, dance and sea, corporeality and nature. This search for a climate aesthetics in Harmattan Theater's work has been a singular but gradual process of cultivating a living practice of climate questions through movement, choreography, dramaturgy, and performance. Cultivating a living praxis of immersion in landscapes and listening to the site for its ecology of movement has been part of the process of constructing and developing durational performances at the juncture of ocean and land.

The name Harmattan suggests the migratory nature of the wind the company is named after. The trans-Saharan Harmattan is a hot dusty orange colored wind that sweeps across the vast Sahara Desert from the eastern coasts of African through the west coast of the continent and all the way to the Americas. The Harmattan wind also known as the Sahara Haze in the United States carries nutrients critical to the ecology of the Amazon rainforests. The gargantuan Sahara Haze traverses the Atlantic from June through October every year, dumping about 20,000 tons of phosphorous in the Amazon rainforests along its blustery transatlantic journey across the seas to the Americas.[1] The Harmattan winds transport detritus and seeds in its blustery transatlantic journey across the Atlantic ocean to the Americas. This trans-Saharan wind is now a distinctive materiality in the North American troposphere, dropping 20,000 tons of Sahara sand across the Atlantic. Manifesting as an orange haze of fine dust traveling light and swiftly through the North American atmosphere, the Harmattan wind is, with climate change, a major part of the climatological transformation of the eastern seaboard of the United States. The Sahara's minerality is part of New York's troposphere from June through October every year. It is this interconnectivity between wind, oceans, materiality of the nonhuman, and the histories of human possibility that Harmattan Theater disinters, disturbs, and dances with. This trans-Saharan wind can occasionally be seen in the Americas as an orange haze of fine dust traveling light and swiftly through the atmosphere: Harmattan.

Toward a Somatic Ecology—Harmattan Performs

By Sofia Varino

Site-specific work requires a willingness to engage with the inhuman and nonhuman histories of a place, engendering the collapse of past and present in a temporal convulsion unfolding into what will unavoidably be an eventual ahuman futurity. How to move through and into this final alterity, how to grapple with the certainty of human extinction while still finding ways to live well with others, accommodating our small daily pleasures alongside incommensurable uncertainty, in creative collaboration, fostering a sociality that is as self-interested as it is generous, curious, responsive?

In setting the terms for this question by pairing these categories, I do not mean to suggest that there is a foundational schism between them or to repeat the dualistic thinking deploying these categories to justify the perpetuation of innumerable violences. Rather, I wish to acknowledge that these dualisms encapsulate much of Western philosophical, scientific, cultural, and political history, and that to consider different categories does not necessarily set them up as opposites but rather linguistically demarcates a spectrum or gradation linking these dialogic terms, blurring their boundaries and emphasizing continuities and ambivalences in how they relate to one another. How to make everyday life more livable, more bearable, for more human and nonhuman communities, on a planetary scale?

At a given site, bodies arrive to begin a movement meditation. Deft hands play musical instruments, filling the air with sound, shaking up the ground with rhythm. A line of bodies begins to take shape, steadily proceeding across the water's edge. Wind blows, sun shines, river flows. Critters and elements riff off each other. One foot in front of another. We carry on. The sky wide open. Harmattan performs.

Note

1 See Temi E. Ologunorisa and Sylvester Keobunah, "The Seasonal Incidence of Sahara Dust Haze over Northern Nigeria," *International Journal of Meteorology* 31, 2006: pp. 271–277.

Bibliography

Abramovich, Marina, *Walk Through Walls: A Memoir*. New York: Crown Archetype, 2016.

Abulafia, David. "Mediterraneans," *Rethinking the Mediterranean*, Ed. Harris W V, Oxford: Oxford University Press, 2005.

Abulafia, David. *The Great Sea: A Human History of the Mediterranean*, Oxford: Oxford University Press, 2013.

Abu-Lughod, J. *Before European Hegemony: The World System A.D. 1250–1350*, New York: Oxford University Press, 1989.

Acri, A (2018) 'The place of Nusantara in the Sanskritic Buddhist cosmopolis', *TRaNS: Trans–Regional and National Studies of Southeast Asia* v6 n2: 139–166.

Ahmed, Sara. *Queer Phenomenology. Orientations, Objects, Others*, Durham: Duke University Press, 2006.

Aikau, Hŏkūlani and Vernadette Gonzalez, "Curating a Decolonial Guide: The Detours Project," *Shima* 13:2 (2019): 11–21.

Unearthing Gotham: *The Archaeology of New York City* by Cantwell, Anne-Marie diZerega and Diana Wall. New Haven: Yale University Press, 2001.

Artaud, Antonin. *The Theater and Its Double*, New York: Grove Press, (1938) 1994.

Bailes, Sara, *Performance Theater and the Poetics of Failure*. London: Routledge, 2010.

Balakrishnan, M and Alexander KM (eds) (1988) *Environmental issues: problems and pollutions*, Trivandrum: Golden Jubilee Publications, University of Kerala.

Banes, Sally. *Greenwich Village 1963: Avant-Garde Performance and the Effervescent Body*. Durham, NC: Duke University Press, 1993.

Barad, Karen, *Meeting the Universe Halfway*. Durham, NC: Duke University Press, 2007.

Bennett, Jane. *Vibrant Matter: A Political Ecology of Things*. Durham, NC: Duke University Press, 2010.

Berlin, Ira, and Leslie M. Harris (eds.). *Slavery in New York*, New York: New Press, 2005.

Bhattacharya, RN (ed) (2001) *Environmental economics: an Indian perspective*, New Delhi: Oxford University Press.

Bishop, Claire. "Antagonism and Relational Aesthetics," *October* 110 (Fall 2004): 51–79.

Bishop, Claire. (ed.) *Participation*. London: Whitechapel Gallery and Cambridge MA: MIT Press, 2006.

Boal, Augusto. *Games For Actors and Non-Actors*, London: Routledge, 1992.

Bourriaud, Nicholas. *Relational Aesthetics*. Trans. Simon Pleasance and Fronza Woods. Dijon: Les Presses du Réel, 2002.

Bracciolini, Poggio. *De l'Inde. Les Voyages en Asie de Niccolo de' Conti: De varietate fortunae livre IV*, Turnhout: Brepols Publishers, 2004.

Braudel, Fernand. *The Structures of Everyday Life: Civilization & Capitalism 15th-18th Century*, Vol. 1, New York: Harper and Row, 1981.

Braudel, Fernand. *The Mediterranean and the Mediterranean World In The Age of Philip II*. Vol. I, Berkeley: University of California Press, 1995.

Braudel, Fernand. *Out of Italy*, New York: Europa Editions, 2019.

Brecht, Bertolt. *Brecht on Theater*. London: Methuen, 1964.

Bush, D, Pilkey Jr. O and Neal, W (1996) *Living by the rules of the sea*, Durham: Duke University Press.

Butler, Judith. (1990) *Gender trouble: Feminism and the Subversion of Identity*. New York: Routledge.

Butler, Judith. *Notes Toward a Performative Theory of Assembly*, Cambridge, MA: Harvard University Press, 2015.

Carson, Rachel. *Silent Spring*, New York: Mariner Books, (1962) 2002.

Carson, Rachel. *The Sea Around Us*, London: Unicorn Press, (1951) 2014.

Cattaneo, Angelo. *Fra Mauro's Mappa Mundi and Fifteenth Century Venice, Terrarum Orbis (TO 8)*, Turnhout: Brepols Publishers, 2011.

Chaudhuri, K.N. *Trade and Civilisation in the Indian Ocean: An Economic History from the Rise of Islam to 1750*, Cambridge: Cambridge University Press, 1985.

Chaudhuri, Una. *Staging Place: The Geography of Modern Drama*. Ann Arbor: University of Michigan Press, 1995.

Chaudhuri, Una and Elinor Fucks (eds). *Land/Scape/Theater*, Ann Arbor: University of Michigan Press, 2002.

Chin, Catherine. "Marvelous Things Heard: On Finding Historical Radiance," *The Massachusetts Review* 58: 3 (2017): 478–491.

Clark, Stephen. "Gaia and the Forms of Life" in *Environmental Philosophy*, edited by Robert Elliot and Arran Gare St. Lucia, Qld: University of Queensland Press, 1983.

Crutzen, Paul J. "Geology of Mankind," *Nature* 415 (January 3, 2002): 23.

Crutzen, Paul J. and E. F. Stoermer. "The 'Anthropocene'," *Global Change Newsletter of the International Geosphere–Biosphere Programme (IGBP)* 41 (May 2000): 17–18.

Das, Gupta Ashin, *India and the Indian Ocean World: Trade and Politics*, New Delhi: Oxford University Press, 2004.

David Harvey, *A Brief History of Neoliberalism*. New York: Oxford University Press, 2005.

Derrida, Jacques. "On Forgiveness," *On Cosmopolitanism and Forgiveness*, translated by Mark Dooley and Michael Hughes, New York: Routledge, 2001: 25–60.

Diedrich, Lisa. *Indirect Action: Schizophrenia, Epilepsy, AIDS, and the Course of Health Activism*. University of Minnesota Press, 2016.

Eco, Umberto. *The Open Work*, Cambridge, MA: Harvard University Press, 1962.

Espinosa, Julio Garcia. "For an Imperfect Cinema". Translated by Julianne Burton. *Jump Cut: A Review of Contemporary Media* 20 ([1979] 2005): 24–26.

Ferreira da Silva, Denise. "Toward a Black Feminist Poethics: The Quest(ion) of Blackness Toward the End of the World," *The Black Scholar*, 44:2 (Summer 2014): 81–97.

Fersuoch, L. *Misreading the Lagoon*, Venice: Corte Del Fontego Editore, 2014.

Finch, R. *The Primal Place*, Woodstock, VT: The Countryman's Press, 1983.

Freire, Paulo. *Pedagogy of the Oppressed*. New York and London: Continuum, (1970) 2006.

Susan L. Glen, *Governors Island*, New York: Arcadia Publishing, 2006.

Guattari, Felix. *Three Ecologies*. London: Continuum, 2000.

Gumbs, Alexis Pauline. *Undrowned: Black Feminist Lessons from Marine Animals*. Chico, CA: AK Press, 2020.

Halberstam, Jack. (2011) *The queer art of failure*. Durham, NC: Duke University Press.

Halberstam, Jack. *Wild Things: The Disorder of Desire*. Durham, NC: Duke University Press, 2020.

Haraway, Donna. *Staying with the Trouble*. Durham, NC: Duke University Press, 2016.

Hartigan, John. "Multispecies vs Anthropocene," *Somatosphere*, December 12, 2014. http://somatosphere.net/2014/multispecies-vs-anthropocene.html/ (last accessed: July 11, 2022).

Hau'ofa, E (1994), 'Our sea of islands', *The Contemporary Pacific* v6 n1: 148–161.

Hayward, Philip. "Aquapelagos and Aquapelagic Assemblages" *Shima: The International Journal of Research into Island Cultures* 6:1 (2012): 1–11.

Henda, Kiluanji Kia. *Plantation – Prosperity and Nightmare* public installation, Lisbon, 2021.

Horden, Peregrine and Nicholas Purcell. 'The Mediterranean and "the new thalassology",' *American Historical Review* 111: 3 (2006): 722–740.

Ingersoll, K.A. *Waves of Knowing: A Seascape Epistemology*, Durham: Duke University Press, 2016.

Jackson, Shannon. *Social Works: Performing Art, Supporting Publics*, New York: Routledge, 2011.

Jackson, Zakiyyah Iman. *Becoming Human: Matter and Meaning in an Antiblack World*, New York: New York University Press, 2020.

Jones, J. W. and Temple, R.C. *The Itinerary of Ludovico di Varthema of Bologna, 1502–1508*, New Delhi: Educational Services, 1997.

Joseph, May. *Fluid New York: Cosmopolitan Urbanism and the Green Imagination*. Durham, NC: Duke University Press, 2013.

Joseph, May. *Sea Log: Indian Ocean To New York*, London: Routledge, 2019a.

Joseph, May. "Kerala Deluge: Archipelagic Thinking," *Overflow*, edited by Andrea Pavoni and Chris Ware. Lo Squandro, 2019b.

Joseph, May. "Indian Ocean Ontology: Nyerere, memory, place," *Reimagining Indian Ocean Worlds*, edited by Smriti Srinivas, Bettina Ng'weno and Neelima Jeychandran. London: Routledge, 2020.

Joseph, May and Macarena Gomez-Barris, "Coloniality and Islands," *Shima* 13:2 (2019): 1–10.

Joseph, May and Sudipta Sen. *Terra Aqua: The Amphibious Lifeworlds of Coastal and Maritime South Asia*. New Delhi: Routledge, 2022.

Joseph, May and Sofia Varino. "Aquapelagic Assemblages: Performing Urban Ecology with Harmattan Theater," *Women's Studies Quarterly* Special Issue: At Sea 45: 1 & 2 (2017): 151–166.

Joseph, May and Sofia Varino. "Multidirectional Thalassology: Comparative Lagoon Ecologies" *Shima*, 15:1 (2021): 256–272.

Kadekodi, G (ed) (2004) *Environmental economics in practice: case studies from India*, New Delhi: Oxford University Press.

Kaplan, E. Ann. *Climate Trauma: Foreseeing the Future in Dystopian Film and Fiction*. New Brunswick, NJ and London: Rutgers University Press, 2015.

Keith, V (2017) *2100: A dystopian utopia—the city after climate change*, New York: Terreform.

Kershaw, Baz. *Theatre Ecology: Environments and Performance Events*. Cambridge: Cambridge University Press, 2007.

Kimmerer, R. (2013). *Braiding Sweetgrass: Indigenous Wisdom, Scientific Knowledge, and the Teachings of Plants*. Minneapolis: Milkweed Editions.

Latour, Bruno. *Facing Gaia: Eight Lectures on the New Climatic Regime*. Cambridge: Polity Press, (2015) 2017.

Latour, Bruno. *Down to Earth: Politics in the New Climatic Regime*. Cambridge: Polity Press, (2017) 2018.

Le Goff, Jacques. *Time, Work & Culture in the Middle Ages*, Chicago: University of Chicago Press, 1980.

Le Goff, Jacques. *Must We Divide History into Periods?* New York: Columbia University Press, 2017.

Lykke, N. (2021) *Vibrant Death: A Posthuman Phenomenology of Mourning*. London and New York: Bloomsbury Academic.

Major, R.H. (ed.), *India in the Fifteenth Century: Being a collection of narratives of voyages to India in the century preceding the Portuguese discovery of the Cape of Good Hope*, London: Routledge, (1857) 2010.

Mancuso, F. *Building on Water*, Venice: Corte Del Fontego Editore, 2014.

Manguin, P.Y. *The Vanishing Jong: Insular Southeast Asian fleets in War and Trade (15th-17th centuries)*, Ithaca: Cornell University Press, 1993.

Manguin, PY, Mani, A and Wade, G (eds) (2011) *Early interactions between South and Southeast Asia: reflections on cross-cultural exchange*, Singapore, New Delhi: Institute of Southeast Asian Studies.

Messeri, Lisa. *Placing Outer Space*, Durham, NC: Duke University Press, 2016.

Mmembe, A. (2019) *Necropolitics*, Durham, NC: Duke University Press.

Neils, Jenifer. *Goddess and Polis: The Panathenaic Festival in Ancient Athens*, Princeton: Princeton University Press, 1993.

Nixon, R. (2011). *Slow Violence and the Environmentalism of the Poor*. Cambridge, Mass.: Harvard University Press.

Nyong'O, Tavia. *Afro-Fabulations: The Queer Drama of Black Life*. New York: New York University Press: 2018.

Ologunorisa, Temi E and Sylvester Keobunah. "The seasonal incidence of Sahara dust haze over northern Nigeria," *International Journal of Meteorology* 31 (2006): 271–277.

Owen D. Gutfreund, "Rebuilding New York In The Auto Age: Robert Moses and his highways". In Hilary Ballon and Kenneth T. Jackson Editors. *Robert Moses and the Modern City: The Transformation of New York*. London and New York: W.W. Norton & Company, Inc, 2007.

Pande, A, Singh, Y, Jasmine, B et al (2014) 'Biodiversity of coastal islands of India' in Sivakumar, K (ed) *ENVIS Bulletin on coastal and marine protected areas in India: Challenges and Way Forward* n15, Dehradun: Wildlife Institute of India: 190-214.

Pereira, W (2007) *Tending the Earth: traditional, sustainable agriculture in India*, Kolkata: Earthcare Books.

Pessoa, Fernando. *Message*, translated by Jonathan Griffith. London: The Menard Press, (1934) 1997.

Phelan, Peggy. *Unmarked. The Politics of Performance*. London and New York: Routledge, 1993.

Polo, Marco. *The Travels of Marco Polo, The Venetian*, New York: E.P. Dutton & Co, (c. 1300) 2003.

Pugh, Jonathan. "Relationality and Island Studies in the Anthropocene," *Island Studies Journal* 13: 2 (2018): 93–110.

Purcell, N. "The Tide, Beach, and Backwash: The Place of Maritime Histories," *The Sea: Thalassography and Historiography*, edited by P.N. Miller. Ann Arbor: University of Michigan Press, 2013.

Rajagopalan, R (2008) *Marine protected areas in India*, Chennai: International Collective in Support of Fishworkers.

Roach, Joseph. *Cities of the Dead: Circum-Atlantic Performance*, New York: Columbia University Press, 1996.

Rothberg, M. *Multidirectional Memory: Remembering the Holocaust in the Age of Decolonization*, Stanford: Stanford University Press, 2009.

Sallée, J (2018) 'Southern Ocean warming', *Oceanography* v31 n2: 52–62.

Salzano, Edoardo. *The Lagoon of Venice: Governance for a Complex System*. Venice: Corte del Fontego Editore, 2014.

Satsuka, Shiho. *Nature in Translation*, Durham, NC: Duke University Press, 2015.

Schechner, Richard. *Environmental Theater*, Hawthorne Books, 1973.

Schneider, Rebecca. *Performing Remains: Art and War in Times of Theatrical Reenactment*. London and New York: Routledge, 2011.

Sengupta, R (2001) *Ecology and economics: an approach to sustainable development*, New Delhi: Oxford University Press.

Settis, Salvatore. *If Venice Dies*, New York: New Vessel Press, 2015.

Sheller, Mimi. *Island Futures. Caribbean Survival in the Anthropocene*, Durham: Duke University Press, 2020.

Srinivas, Smriti, Bettina Ng'weno and Neelima Jeychandran, Eds. *Reimagining Indian Ocean Worlds*, London: Routledge, 2020.

Subrahmanyam, Sanjay, *The Political Economy of Commerce: Southern India 1500–1650*, New Delhi: Cambridge University Press, 2004.

Thiong'O, Ngũgĩ Wa. *Decolonizing the Mind: the Politics of Language in African Literature*, Heinemann Educational, 1986.

Thompson, Nato and Independent Curators International. *Experimental Geography: Radical Approaches to Landscape, Cartography, and Urbanism*, Brooklyn NY: Melville House, 2008.

Vadakkekara, Benedict. *Origin of Christianity in India: A Historiographical Critique*, New Delhi: Media House, 2007.

Varino, Sofia. *Vital Differences. Indeterminacy and the Biomedical Body* (Ph.D. Dissertation), Stony Brook: Stony Brook State University of New York, 2017.

Varino, Sofia. "Pathogenic (Auto)Ecologies: Environmental Illness Mechanisms." *The Bloomsbury Handbook to the Medical-Environmental Humanities*, edited by Scott Slovic, Swarnalatha Rangarajan, Vidya Sarveswaran. London and New York: *Bloomsbury*, 2022.

Varthema, L (1997) *The itinerary of Ludovico di Varthema of Bologna from 1502 to 1508* (translated by Jones, JW), New Delhi, Madras: Asian Educational Services.

Venkataraman, K and Wafar, M (2005) 'Coastal and marine biodiversity of India', *Indian Journal of Marine Sciences* v34 n1: 57–75.

Paulo Virno, *A Grammar of the Multitude: For an Analysis of Contemporary Forms of Life*. Translated by Isabella Bertoletti, James Cascaito and Andrea Casson. Cambridge: M.I.T. Press, 2004.

Weber, Andreas. *Enlivenment: Towards a Fundamental Shift in the Concepts of Nature, Culture and Politics*. Berlin: Heinrich Böll Stiftung, 2013.

Woodward, HW (2004) 'Review article: esoteric Buddhism in Southeast Asia in the light of recent scholarship', *Journal of Southeast Asian Studies* v35 n2: 329–354.

Yusoff, Kathryn. *A Billion Black Anthropocenes or None*. Minneapolis, MN: University of Minnesota Press, 2018.

Index

Page numbers followed by n refer notes.

For Product Safety Concerns and Information please contact our EU representative GPSR@taylorandfrancis.com
Taylor & Francis Verlag GmbH, Kaufingerstraße 24, 80331 München, Germany

www.ingramcontent.com/pod-product-compliance
Lightning Source LLC
LaVergne TN
LVHW010933110826
845149LV00013B/2585

* 9 7 8 1 0 3 2 4 1 8 2 6 1 *